THE RINGS OF DIMENSION TWO

PURE AND APPLIED MATHEMATICS

A Program of Monographs, Textbooks, and Lecture Notes

LECTURE NOTES
IN PURE AND APPLIED MATHEMATICS

1. *N. Jacobson,* Exceptional Lie Algebras
2. *L.-Å. Lindahl and F. Poulsen,* Thin Sets in Harmonic Analysis
3. *I. Satake,* Classification Theory of Semi-Simple Algebraic Groups
4. *F. Hirzebruch, W. D. Newmann, and S. S. Koh,* Differentiable Manifolds and Quadratic Forms
5. *I. Chavel,* Riemannian Symmetric Spaces of Rank One
6. *R. B. Burckel,* Characterization of C(X) Among Its Subalgebras
7. *B. R. McDonald, A. R. Magid, and K. C. Smith,* Ring Theory: Proceedings of the Oklahoma Conference
8. *Yum-Tong Siu,* Techniques of Extension of Analytic Objects
9. *S. R. Caradus, W. E. Pfaffenberger, and Bertram Yood,* Calkin Algebras and Algebras of Operators on Banach Spaces
10. *Emilio O. Roxin, Pan-Tai Liu, and Robert L. Sternberg,* Differential Games and Control Theory
11. *Morris Orzech and Charles Small,* The Brauer Group of Commutative Rings
12. *S. Thomeier,* Topology and Its Applications
13. *Jorge M. López and Kenneth A. Ross,* Sidon Sets
14. *W. W. Comfort and S. Negrepontis,* Continuous Pseudometrics
15. *Kelly McKennon and Jack M. Robertson,* Locally Convex Spaces
16. *M. Carmeli and S. Malin,* Representations of the Rotation and Lorentz Groups: An Introduction
17. *George B. Seligman,* Rational Methods in Lie Algebras
18. *Djairo Guedes de Figueiredo,* Functional Analysis: Proceedings of the Brazilian Mathematical Society Symposium
19. *Lamberto Cesari, Rangachary Kannan, and Jerry D. Schuur,* Nonlinear Functional Analysis of Differential Equations: Proceedings of the Michigan State University Conference
20. *Juan Jorge Schäffer,* Geometry of Spheres in Normed Spaces
21. *Kentaro Yano and Masahiro Kon,* Anti-Invariant Submanifolds
22. *Wolmer V. Vasconcelos,* The Rings of Dimension Two

Other volumes in preparation

THE RINGS OF DIMENSION TWO

Wolmer V. Vasconcelos

Department of Mathematics
Rutgers University
The State University of New Jersey
New Brunswick, New Jersey

MARCEL DEKKER, INC. NEW YORK AND BASEL

MARCEL DEKKER, INC.

270 Madison Avenue, New York, New York 10016

LIBRARY OF CONGRESS CATALOG CARD NUMBER: 76-592

ISBN: 0-8247-6447-1

Current printing (last digit):
10 9 8 7 6 5 4 3 2 1

PRINTED IN THE UNITED STATES OF AMERICA

To Aurea

INTRODUCTION

A motivating element in these notes on homological dimensions of commutative rings is the following problem: Let A be a ring, and let E be a finitely generated A-module with generators $f_1, \ldots, f_n$. The relations (or syzygies) of the f_i's are the n-tuples $(g_1,\ldots,g_n) \in A^n$ such that

$$g_1f_1 + \cdots + g_nf_n = 0$$

They are made into an A-module K which, if finitely generated, allows one to "find" the relations among a finite set of generators. If this procedure can be carried out a certain number of times, we get a resolution:

$$A^{m_r} \longrightarrow \cdots \longrightarrow A^{m_1} \longrightarrow A^{m_0} \longrightarrow E \longrightarrow 0$$

The problem is whether the integer r is sufficiently large to ensure that from the rth stage the resolution can be indefinitely continued using finitely generated modules. In the language of Ref. 11, Chap. 1, when do modules with a finite r-presentation have a finite infinite presentation?

To formalize this question, we define the λ-dimension of A [λ-dim(A) for short] as the infimum of such integers--if any exists. Thus Noetherian rings are those with λ-dimension 0, while the rings of λ-dimension at most 1, named coherent, are those for which finitely generated ideals are finitely presented. Note that, according to Weyl [65], before succeeding in proving his 'basis theorem' Hilbert in fact proved the coherence of the polynomial ring in several indeterminates over a field.

The question of determining the λ-dimension of a ring becomes simpler if we were to know that projective resolutions stop after a certain number of steps; such rings will be called "regular," and its (grandiosely named) global dimension [gl dim(A) for short] is the least number of steps that works for all modules. Although the λ-dimensions of a ring A and a polynomial ring $B = A[x_1,\ldots,x_n]$ may differ, a measure of stability is expressed in the inequality $\lambda\text{-dim}(B) \leq 1 + \text{gl dim}(A)$. One of the main results in these notes is the proof that if A is coherent and of global dimension two, then B above is also coherent. A feature of the proof is, once sufficiently many structural details of A are revealed, to throw in all indeterminates simultaneously--in marked contrast to the Hilbert's basis theorem--to keep better track of the coefficients.

The bulk of the notes is dedicated to a leisurely discussion and assemblage of the pieces of homological dimension theory needed for a description of the rings of global dimension two.

Very early it became clear that the classical examples of such rings, Noetherian rings and Prüfer rings of appropriate size, were being used with singular frequency. Thus, the running theme became the examination of the extent with which such rings make up all rings of global dimension two. The most pleasing aspect of this program concerned the local rings; they were shown to be either Noetherian rings, valuation domains, or a hybrid obtained by attaching one of the former class on the top of a valuation ring. Guided by this view, the next step was the examination of the domains of global dimension two; here the results were far from complete, perhaps because of the diverse ways such rings arise, as illustrated by the examples scattered throughout. Noteworthy is the result that a domain which is not a Prüfer ring must contain a finitely generated maximal ideal. This was, however, as far as the analysis proceeded except for the examination of its more immediate consequences. Finally, for arbitrary rings (*i.e.*, not necessarily domains) it is clear that at least for "half" of them, the coherent rings, the difficulties are just the same as for domains. In this case a promising sheaf representation over a compact space exists with fibers which are domains of dimension at most two. The noncoherent rings were just identified with the inexcusable purpose of steering away from them.

Every chapter contains a description of its contents. An overall summary is as follows: Chapter 0 is a brief survey of homological dimension theory dealing mostly with the relationship among the following dimensions associated to a ring: Krull, global, weak, and finitistic. Chapter 1 gives descriptions of rings in which the last three dimensions assume the value one; this is needed later since such rings appear as rings of residues or fractions of rings of dimension two. Chapters 2 and 4 contain the results on local rings. Chapters 3 and 5 are of less specialized interest since the first is concerned with descent of projectivity, and the other with change of rings and dimensions. A consequence of Chapter 3 is that it allows, in a rather painless way, construction of local rings of finite global dimension but with many zero divisors. Chapter 6 exploits the local theorem and change of rings to treat the nature of the ring modulo a finitely generated ideal. Chapter 7 deals mostly with domains in terms of infinitely many local and lower dimensional residual conditions. Chapter 8 uses the accumulated information to show the coherence of the polynomial ring over a coherent ring of global dimension two. It concludes with a discussion of the λ-dimension of rings.

The author is grateful to the lively audience of a seminar held at Rutgers University in the spring of 1973, where the subject was discussed, and especially to Brian Greenberg and Hu Sheng, both of whom spoke of their own work (Chapters 4 and 6, respectively). Greenberg has also played a role in the current form of Chapter 8.

We are equally indebted to Laurent Gruson and Christian Jensen for a spirited correspondence that we, at times (rather shamelessly!), transcribe freely into the text.

Finally, we thank Irving Kaplansky who, so many years ago, made this subject attractive to us and continued to show interest in its development.

A work about the bibliography: Although plentiful, it is certainly not exhaustive and no exaggerated care was taken in providing direct attribution of all the results contained here.

Wolmer V. Vasconcelos

CONTENTS

THE RINGS OF DIMENSION TWO

Chapter 0

PRELIMINARIES

All rings to be considered are commutative--and this will be emphasized occasionally--and have an identity; modules are unital. References 11 and 36 should be used for terminology and basic notions of commutative algebra. In this chapter we start assemblying various results in the theory of homological dimensions to be repeatedly referred to. To set the stage, we begin with the definitions of our main objects of interest.

1. *Global dimensions*

0.1 *Definition.* Let E be a module over the ring A; the projective dimension of E ($pd_A E$ for short or simply pd E when no danger of confusion exists) is finite and equal to n if there is an exact sequence

$$0 \longrightarrow P_n \longrightarrow \cdots \longrightarrow P_1 \longrightarrow P_0 \longrightarrow E \longrightarrow 0$$

with P_i A-projective and n least. Otherwise the projective dimension of E is said to be infinite.

0.2 *Definition.* Global dimension of A = gl dim(A) = sup{pd E}, over all A-modules.

The following result conveniently reduces the task of determining the global dimension of a ring A [2].

0.3 *Lemma* (Global dimension lemma). gl dim(A) = sup pd{A/I}, where I runs over the ideals of A.

Proof: Assume sup pd A/I $= n < \infty$, and let E be an A-module for which we must show pd E $< n + 1$, that is $\mathrm{Ext}_A^{n+1}(E,F) = 0$, for any A-module F. Consider an injective presentation of F,

$$0 \longrightarrow F \longrightarrow I_0 \longrightarrow I_1 \longrightarrow \cdots \longrightarrow I_{n-1} \longrightarrow C \longrightarrow 0$$

where the I_j's are injective modules. Since $\mathrm{Ext}_A^1(E,C) = \mathrm{Ext}_A^{n+1}(E,F)$, it suffices to show that C is injective. This will be done, by Baer's lemma [14], if $\mathrm{Ext}_A^1(A/I,C) = 0$ for any ideal I. But this last module is isomorphic to $\operatorname{Ext}_A^{n+1}(A/I,F) = 0$.

0.4 *Remark.* If A is a Noetherian ring, it is enough to consider the ideals generated by three elements [12,27,38], or more efficiently maximal ideals, which reduces the task in the case of a local ring to the examination of a single ideal [3,56].

0.5 *Definition.* Let E be an A-module; E is said to be A-flat (or simply flat) if for an arbitrary exact sequence of A-modules,

$$(M) \qquad \cdots M_{i+1} \longrightarrow M_i \longrightarrow M_{i-1} \cdots$$

the sequence

$$(E \otimes_A M) \qquad \cdots E \otimes_A M_{i+1} \longrightarrow E \otimes_A M_i \longrightarrow E \otimes_A M_{i-1} \cdots$$

is also exact.

Since the functor $E \otimes_A (-)$ already preserves right exactness, the flatness of E amounts to the fact that $E \otimes_A (-)$ preserves injections.

If $\underline{a} = \{a_1, \ldots, a_n\}$ is a sequence of elements of A, the module of relations on the a_i's is the kernel of

$$\phi\colon A^n \longrightarrow A, \qquad \phi(x_1, \ldots, x_n) = \Sigma x_i a_i$$

$R_E(\underline{a}) = \ker(1_E \otimes \phi)$ is then $\{(e_1, \ldots, e_n) \in E^n$ with $\Sigma a_i e_i = 0\}$; it will be called the module of relations on the a_i's with coefficients in E. Notice that $R_A(\underline{a})E \subset R_E(\underline{a})$.

0.6 *Proposition.* The following are equivalent for an A-module E.

(1) E is flat

(2) For every sequence $\underline{a}$ as above, $R_A(\underline{a})E = R_E(\underline{a})$.

Proof: (1) $\Rightarrow$ (2): Let $(\underline{a})$ be the ideal generated by the elements in $\underline{a}$. Consider the exact sequence

$$F \xrightarrow{\psi} A^n \xrightarrow{\phi} A \longrightarrow A/(\underline{a}) \longrightarrow 0$$

where F is a free module. Tensoring with E, we obtain the exact sequence

$$E \otimes F \xrightarrow{1 \otimes \psi} E \otimes A^n \xrightarrow{1 \otimes \phi} E \otimes A \longrightarrow E \otimes A/(\underline{a}) \longrightarrow 0$$

(where the A's were dropped harmlessly in $\otimes_A$). Since $\operatorname{im}(1 \otimes \psi) = R_A(\underline{a})E$, we have (2). Actually, in the language of derived functors, the argument shows that

$$\operatorname{Tor}_1^A(A/(\underline{a}),E) = R_E(\underline{a})/R_A(\underline{a})E$$

(2) $\Rightarrow$ (1): Consider the diagram

$$\begin{array}{ccc} & & P \\ & & \downarrow \psi \\ F & \xrightarrow{\phi} & G \end{array}$$

where ϕ is injective, P is a free module, and ψ is surjective. With $L = \ker \psi$ and $K = \psi^{-1}(\phi(F))$, we obtain the diagram

$$\begin{array}{ccccccc} E \otimes L & \longrightarrow & E \otimes P & \longrightarrow & E \otimes G & \longrightarrow & 0 \\ \uparrow & & \uparrow & & \uparrow & & \\ E \otimes L & \longrightarrow & E \otimes K & \longrightarrow & E \otimes F & \longrightarrow & 0 \end{array}$$

with obvious maps. From this it follows that the vertical map on the right will be injective if the two maps ending in $E \otimes P$ are injective. We may then assume that G is a free module. It is also clear that G may be assumed to be of finite rank. An easy induction on rank takes care of the remainder of the proof, the starting case being supplied by the hypothesis.

Let E be a flat A-module, and consider a presentation

$$0 \longrightarrow K \xrightarrow{\psi} F \xrightarrow{\phi} E \longrightarrow 0$$

where F is free with basis $\{f_\alpha\}$.

If $k = \Sigma a_i f_{\alpha_i}$ is an element of K, tensoring this sequence by $A/(\underline{a})$, $\underline{a} = \{a_1, \ldots, a_n\}$, yields by (0.6) an exact sequence and $k \in (\underline{a})F$. Thus $k = \Sigma a_i g_i$, $g_i \in K$. If we define $\theta\colon F \to K$ by $\theta(f_{\alpha_i}) = g_i$ and arbitrarily on the rest of the basis, we get a homomorphism such that $\theta(k) = k$. More generally, we have the following proposition.

<u>0.7 *Proposition*</u>. If $k_1, \ldots, k_n$ are elements of K, there exists a homomorphism $\theta\colon F \to K$ fixing all k_i's.

<u>*Proof:*</u> We use an argument of S. U. Chase. Taking care of the case $n = 1$ by the remarks above, we proceed by induction on n. Let $\theta_n\colon F \to K$ be such that $\theta_n(k_n) = k_n$. Let $v_i = k_i - \theta_n(k_i)$. By induction there exists a homomorphism $\theta'\colon F \to K$ such that $\theta'(v_i) = v_i$ for all v_i's. Define

$$\theta = I - (I - \theta')(I - \theta_n)$$

It is easy to check that θ has the desired properties.

<u>0.8 *Definition*</u>. Let E be an A-module; the <u>flat dimension</u> of E (flat dim E for short) is finite and equal to n if there is an exact sequence

$$0 \longrightarrow F_n \longrightarrow \cdots \longrightarrow F_1 \longrightarrow F_0 \longrightarrow E \longrightarrow 0$$

with F_i A-flat and n least.

0.9 *Definition.* Weak dimension of A = weak dim(A) = sup{flat dim E} over all A-modules.

For diversity, and euphony, we shall refer to the weak dimension of A as the Tor-dimension of A an abbreviate it Tor-dim(A). The result corre- ponding to (0.3) assumes here the particularly simple form of the following lemma

0.10 *Lemma.* Tor-dim(A) = sup{flat dim A/I}, I running over the finitely generated ideals of A.

A comparison of (0.3) to (0.10) shows that the requirement that a ring A have finite global dimension is much stronger than the requirement that it have finite weak dimension. To help bridge the gap, the following (and its generalizations) is a key tool [29].

0.11 *Lemma* (Jensen's lemma). If

$$0 \longrightarrow K \longrightarrow P \xrightarrow{\phi} E \longrightarrow 0$$

is an exact sequence where E is flat, P is projective, and both P and K are countably generated, then K is projective.

Proof: Assume that P is free on e_i, i = 1, 2, ... If C is a finitely generated submodule of K, say, on $c_1, \ldots, c_\mu$, by (0.7) there is a homomorphism $u: P \to K$ such that $u(c_i) = c_i$, for all c_i's.

If $c_i = \Sigma r_{ij} e_j$, the elements $d_j = u(e_j)$ satisfy the relations $c_i = \Sigma r_{ij} d_j$. Let u_C be the endomorphism of P defined by

$$u_C(e_j) = e_j - d_j, \text{ if } e_j \text{ occurs in the } c_i\text{'s}$$

$$u_C(e_j) = e_j, \text{ if otherwise}$$

We then have $u_C(c) = 0$ for all elements of C. The module generated by the d_j's will be denoted by C^*.

Since K is countably generated, it may be written as an ascending union of finitely generated submodules K_n:

$$K = \cup K_n, \qquad K_1 \subset K_2 \subset \cdots \subset K_n \subset \cdots$$

Define inductively

$$C_1 = K_1, \ C_2 = C_1^* + K_2, \ \ldots, \ C_n = C_{n-1}^* + K_n$$

Let u_n denote the endomorphism obtained when the above construction is applied to the module C_n. The following relations are easily checked:

(1) $\phi u_n = \phi$

(2) $u_n(c) = 0$ for $c \in C_n$

(3) $u_n u_m = u_n$ for $n > m$

Let F be the direct sum of a countable set of copies of P. We claim that the sequence

$$0 \longrightarrow F \xrightarrow{\chi} F \xrightarrow{\psi} E \longrightarrow 0$$

is exact, with

$$\psi(x_1,x_2,\ldots) = \phi(\Sigma x_i)$$

and

$$\chi(x_1,x_2,\ldots) = (x_1,\ x_2 - u_1(x_1),\ \ldots,\ x_n - u_{n-1}(x_{n-1}),\ \ldots)$$

It is clear that ψ is surjective and χ is injective. From (1), im χ ker ψ. For the converse, let $(x_1,x_2,\ldots)$ be an element of ker ψ, that is, Σx_i belongs to $\ker \phi = K$, $x_i \in C_n$ for a suitable C_n. Then (2) implies $u_n(\Sigma x_i) = 0$. By (3) we may assume $x_i = 0$ for $i > n$. By repeated use of (3), we check that

$$\chi(x_1,\ u_1(x_1) + x_2,\ u_2(x_1 + x_2) + x_3,\ \ldots) = (x_1,x_2,\ldots)$$

to conclude the proof.

The following definition is another side of the global dimension of a ring.

0.12 *Definition.* The finitistic projective dimension of A is defined as in (0.2) by taking only the modules with finite projective dimension. We abbreviate it as FPD(A).

This dimension together with the Tor-dimension of A are indeed the two faces of the global dimension [32].

0.13 *Theorem.* Let A be a ring for which FPD(A) is finite. Then any flat module has finite projective dimension.

In particular, the finiteness of gl dim(A) is a consequence of the simultaneous finiteness of FPD(A) and Tor-dim(A). In general, these three numbers are distinct, with the only relationship being contained in the formula gl dim(A) = sup{Tor-dim(A), FPD(A)}.

With d(A) denoting any of the three dimensions above [35], a way to produce rings with finite dimensions is provided by the following classical result.

0.14 *Theorem* (Hilbert syzygies' theorem). Let $A[x]$ denote the ring of polynomials in one indeterminate over A. Then $d(A[x]) = d(A) + 1$.

Proof: (This proof uses a device introduced by Hochschild, together with some changes of rings and homological dimensions results [35].)

Let E be an $A[x]$-module; construct the sequence

$$0 \longrightarrow A[x] \otimes_A E \xrightarrow{\psi} A[x] \otimes_A E \xrightarrow{\phi} E \longrightarrow 0$$

where

$$\psi(x^n \otimes e) = x^n \otimes xe - x^{n+1} \otimes e$$

$$\phi(f(x) \otimes e) = f(x)e$$

It is easy to see that this sequence of $A[x]$-modules and $A[x]$-homomorphisms is exact. The proof proceeds now as follows: If E is an A-module, it can be viewed as an $A[x]$-module via $f(x) = f(0)e$. The change of rings result of Ref. 35 then says that $\mathrm{pd}_{A[x]}E = \mathrm{pd}_A E + 1$.

If now E is an $A[x]$-module, the sequence above says that $\mathrm{pd}_{A[x]}E \geq \mathrm{pd}_A E \geq \mathrm{pd}_{A[x]}A[x] \otimes E$. The first inequality follows because $A[x]$ is A-free. Thus $\mathrm{pd}_{A[x]}E \leq \mathrm{pd}_A E + 1$.

By taking sup over either all modules or only those of finite projective dimension, we prove the syzygies' theorem for gl dim(A) or FPD(A). The case of Tor-dim(A) is similar.

2. *Noetherian rings*

We recall the definition of a regular (Noetherian) local ring. Let A be a local ring of maximal ideal $\underline{m}$. If the Krull dimension of A equals the dimension of the $A/\underline{m}$-vector space $\underline{m}/\underline{m}^2$, A is said to be regular. This remarkable class of rings is characterized homologically in the following theorem [3,56].

0.15 *Theorem* (Auslander-Buchsbaum-Serre theorem). Let A be a Noetherian ring. Then gl dim(A) is finite if and only if A has finite Krull dimension and the localization $A_{\underline{m}}$ is regular for each maximal ideal $\underline{m}$.

For a commutative Noetherian ring the global and the weak dimensions coincide according to (0.3) and (0.10). Of much more recent vintage is the determination of the finitistic dimension of a Noetherian ring A [26].

0.16 *Theorem.* FPD(A) = Krull dimension of A.

The surprising fact is that this equality had not been proved even for low-dimensional cases other than for Artinian rings, for which it is trivial, or for regular rings as in (0.15).

3. *Dimension zero*

It is not particularly difficulty to "identify" the rings for which the dimensions defined above are zero. Thus we have the following.

0.17 gl dim(A) = 0 means $A = K_1 \times \cdots \times K_n$, K_i = field

Indeed, gl dim(A) = 0 simply means that all modules are projective, that is, A is a semisimple (commutative) ring.

0.18 Tor-dim(A) = 0 means that A is a von Neumann ring, that is, every ideal Aa is generated by one idempotent.

To see this, let I be a finitely generated ideal and consider the sequence

$$0 \longrightarrow I \longrightarrow A \longrightarrow A/I \longrightarrow 0$$

Since A/I is flat, by (0.7) there is a homomorphism $u: A \to I$ fixing a generating set for I. Thus the sequence splits and I is generated by one idempotent.

0.19 [6]. FPD(A) = 0 means that $A = B_1 \times \cdots \times B_n$, where each B_i is a ring with a unique prime ideal P_i which is T-nilpotent (*i.e.*, if we pick a sequence $a_1, a_2, \ldots$ of elements in P, then for some m, $a_1 \cdot a_2 \cdots a_m$ is zero).

Indeed assume FPD(A) = 0 and let $a_1, a_2, \ldots$ be a sequence of elements in A. Construct the A-homomorphism $\phi: A[x] \to A[x]$ where $\phi(x^n) = x^n - a_{n+1}x^{n+1}$. ϕ is clearly injective and $E = \text{coker } \phi$ has projective dimension at most one. The assume E to be projective; note that $E = J \cdot E$, where J is the ideal generated by the a_i's.

Write $E \oplus G \simeq A[x]$, and let ψ be the projective of A[x] onto E. If $e \in E$, write

$$e = b_0 + b_1x + \cdots + b_nx^n$$

and denote $c(e) = (b_0, \ldots, b_n)$. Applying ψ to the equation above, one gets $e = \psi(e)$, and since $E = J \cdot E$, we conclude $c(e) = J \cdot c(e)$. By Nakayama's lemma we can find $j \in J$ with $(1 - j)c(e) = 0$.

We apply these remarks to the class of 1 in E. Since $(1 - j)c(e) = 0$ we conclude $(1 - j)e = 0$ also. $1 - j$ is then in the image of ϕ,

$$1 - j = b_0(1 - a_1x) + \cdots + b_n(x^n - a_{n+1}x^{n+1})$$

an easy calculation now shows

$$(1 - j)a_1 \cdot a_2 \cdots a_{n+1} = 0$$

We are now in a position to prove (0.19). Let us begin by showing that A is zero-dimensional, that is, every prime ideal is maximal. If not, pick $a \in \underline{m} \setminus \underline{p}$, $\underline{p} \subset \underline{m}$ distinct prime ideals. With $a = a_1 = a_2 = \cdots$ above we get for some $r \in A$,

$$(1 - ra)a^{n+1} = 0$$

and $1 - ra \in \underline{p} \subset \underline{m}$, a contradiction.

Next we show that A is a semilocal ring. Assume not. Since the ring B = A/(nil radical) is a von Neumann ring, we can find an infinite sequence of orthogonal idempotents $e_1^*, e_2^*, \ldots$ in B. Lift them to A (but cheaply; there is no need to assume $e_1, e_2, \ldots$ are orthogonal). Notice that $I = (e_1, e_2, \ldots)$ is a flat ideal of A. By (0.11), A/I has projective dimension at most one, and I will be generated by a single element.

Thus we can write $A = B_1 \times \cdots \times B_n$, B_i = local. Let P_i be the unique prime of B_i. Since $\mathrm{FPD}(B_i) = 0$ also, the argument above shows that P_i is T-nilpotent.

The converse is immediate.

Remark. The last result is already an indication on how much simpler the determination of these dimensions for commutative rings is compared with the noncommutative cases. For another instance, in Chapter 1 we shall see how painless it is to give a reasonable description of the commutative rings of global dimension one, whereas for noncommutative rings the similar situation is not totally satisfactory. In higher dimensions the noncommutative case is chaotic, to put it mildly.

Chapter 1

DIMENSION ONE

Here we discuss briefly what it means for a commutative ring to have Tor-dimension, global dimension or finitistic projective dimension one. The need for such a sketch will become apparent later as those properties will show up in rings of fractions or residues of rings of dimension two. This chapter concludes with a discussion on generating efficiently finitely generated ideals in rings of dimension one.

1. Tor-dimension one

1.1 *Lemma.* A finitely generated flat module over a local ring is free.

Proof: Let E be a finitely generated flat module over the local ring $\{A,\underline{m}\}$. Among all the minimal generating sets for E pick one $e_1, \ldots, e_n$ with a relation

$$r_1e_1 + \cdots + r_se_s = 0 \qquad r_i \neq 0,\ s \leq n$$

of shortest length. From the flatness of E we may write

$$e_i = \Sigma a_{ij}f_j \qquad \Sigma r_i a_{ij} = 0$$

With all $a_{ij} \in \underline{m}$ we would have

$$E = \underline{m}.E + (e_{s+1}, \ldots, e_n)$$

and thus $s = 0$, by Nakayama's lemma. Assume then $s > 0$ and let, say, $a_{1,1} \notin \underline{m}$. But in this case we have a relation

$$r_2(e_2 - a_{1,1}^{-1} \cdot a_{2,1}e_1) + \cdots + r_s(e_s - a_{1,1}^{-1} \cdot a_{s,1}e_1) = 0$$

which is shorter than the one above, in terms of the minimal generating set $\{e_1, e_2 - a_{1,1}^{-1} \cdot a_{2,1}e_1, \ldots, e_s - a_{1,1}^{-1} \cdot a_{s,1}e_1, e_{s+1}, \ldots, e_n\}$.

1.2 *Proposition.* A has Tor-dimension ≤ 1 if and only if A_P is a valuation domain for each prime ideal P.

Proof: Lemma 1.1 says that every finitely generated ideal of a local ring V of Tor-dim ≤ 1 is free and thus V is a valuation domain. As the flatness is decided at the localizations, the statement follows.

1.3 *Examples.* (a) Let A be the ring of all algebraic integers; then since every finitely generated ideal of A is principal [36], A has Tor-dimension one. As A is a countable ring, it follows from Jensen's lemma (0.11) that every ideal has projective dimension at most one and thus global dim(A) ≤ 2. (It is clearly not one, as will be seen shortly.)

(b) Let E be a countable direct sum of copies of $\mathbb{Z}/2\mathbb{Z}$ with addition and multiplication defined component-wise. Let $A = \mathbb{Z} \oplus E$, and define a multiplication according to the rule $(m,e) \cdot (n,f) = (me, mf + ne + ef)$. It follows easily that for each prime ideal P of A, A_P is either $\mathbb{Z}/2\mathbb{Z}$ or $\mathbb{Z}_{(p)}$. Thus A is still another ring of Tor-dimension one and, as we shall see, of global dimension two.

2. *Global dimension one*

Our approach to a description of such rings, *i.e.* of hereditary rings, will be through the examination of the conditions that make a projective ideal finitely generated.

First we recall the notion of trace of a projective module E over the commutative ring A. It is simply the ideal $J = J(E) = \Sigma f(E)$, where f runs over $\mathrm{Hom}_A(E,A)$. Equivalently, J is the ideal generated by the coordinates of all elements of E whenever a decomposition $E \quad G = F$ (free) is given. From this second interpretation it follows that if $h\colon A \to B$ is a ring homomorphism, then $J(E \otimes B) = h(J(E))B$.

At this point we insert, for quick reference, the following result which will be indispensable throughout [34].

1.4 *Theorem.* A projective module E over a local ring is free.

1.5 *Proposition.* Let J be the trace of a projective module. If J is finitely generated, then $J = Ae$, e an idempotent.

Proof: Let J be the trace of E and let P be a prime ideal of A. If $E_P = 0$, $J_P = 0$ by the remarks above, whereas if $E_P \neq 0$, (1.4) says that E_P is free and thus $J_P = A_P$. In particular, $J = J^2$ (checked at the localizations). If J is finitely generated, the so-called determinantal trick then yields the desired conclusion.

Note that in all cases A/J is a flat A-module. Such ideals will be referred to as pure ideals.

1.6 *Proposition.* Let I be a projective ideal of the commutative ring A. If I is not contained in any minimal prime ideal, it is finitely generated.

Proof: Let J be the trace ideal of I. If J = A, it would mean that we have an equation

$$1 = \sum_1^n f_i(x_i) \qquad f_i \in \mathrm{Hom}_A(I,A),\ x_i \in I$$

Thus for $x \in I$ we have

$$x = \sum_1^n x f_i(x_i) = \sum_1^n f_i(x) x_i$$

and the x_i's generated I. We show this to be the case for J. Notice that $I \subset J$; if J is not A, let M be a maximal ideal containing it. By hypothesis M is not a minimal prime ideal and will thus properly contain one such, say, P. Consider the canonical homomorphism $A \to A/P$; by the remarks above, $I \otimes_A A/P = I/PI$ is a projective A/P-module of trace $(J + P)/P$ M/P. But an integral domain (viz. A/P) cannot admit a pure ideal [viz. $(J + P)/P$] other than the trivial ones, and this is ruled out here. This contradiction proves the statement.

The first use to be made of (1.6) is under the most favorable conditions: study of the reduced rings (*i.e.*, rings without nilpotent elements) in which every ideal not contained in any minimal prime ideal is projective. (Notice that if A has global dimension one, every localization of it is a domain; thus it is reduced). For brevity we call such rings almost hereditary.

1.7 *Theorem.* Let A be almost hereditary. Then:

(1) A is semihereditary (*i.e.*, every finitely generated ideal is projective)

(2) A/P is a Dedekind domain for each prime ideal P

(3) A/I is Artinian for every ideal I not contained in any minimal prime, and I is a product of maximal ideals.

Proof: The statements will follow from a list of observations on A.

(a) For every prime ideal M, the localization A_M is an integral domain. Since A is reduced, it is enough to show that the minimal primes are two by two comaximal. Consider P, Q distinct minimal primes, and suppose $P + Q \subset M$, a maximal ideal. By hypothesis $P + Q$ is a projective ideal; in A_M both P and Q survive and $P_M + Q_M$ is generated by one element $x + y$, $x \in P_M$, $y \in Q_M$. Now $x = z(x + y)$ for some $z \in A_M$; if z is a unit, $y \in P_M$ and $Q_M \subset P_M$; if z is not a unit, $(1 - z)x = zy$ gives

$x \in Q_M$ and $P_M \subset Q_M$. Either way we get a contradiction.

(b) The annihilator of each element of A is generated by one idempotent. Let $x \in A$ and write I = annihilator of x. By (a), for each prime M, $I_M = 0$ or A_M and in particular $I = I^2$. Let $J = (x,I)$; since I is a pure ideal, $I \cap (x) = Ix = 0$ and $J = (x) \oplus I$. On the other hand, if J is contained in a minimal prime P, A_P is a field and the image of x in A_P is trivial; consequently, $I_P = A_P$, which is impossible. Thus J is a projective ideal, and according to (1.6) it is finitely generated. Then I is also finitely generated and since $I = I^2$, $I = Ae$ for some idempotent element e.

Now we prove (1): Let $I = (x_1, \ldots, x_n)$ be a finitely generated ideal which must be proved projective. If Ae_i = annihilator of x_i, e_i = idempotent, then $Ae_1 \cap \ldots \cap Ae_n = Ae$ $(e = e_1 \cdots e_n)$ is the annihilator of I. Let $J = (e,I)$; as before we conclude that $J = Ae \oplus I$ and that J is not contained in a minimal prime. Hence it is projective, and I is also projective.

As for (2), (a) says that each minimal prime P is a pure ideal and thus A/P is a flat epimorphic image of A. It obviously inherits the defining property of A and being a domain, it is a Dedekind domain.

Finally for (3), if I is as stated, A/I is a Noetherian ring by (1.6), and since A has Krull dimension one by (2), there are only finitely many maximal ideals containing I; since each localization A_M is a discrete valuation ring, it follows that I is a product of maximal ideals.

In gneral, the remaining properties of A are hidden in the nature of some of its associated rings, *e.g.*, total ring of quotients, Boolean ring of idempotents. Finally, we discuss a phenomenon--studied in Ref.43 by valuation theory and in Ref.8 by sheaf theory--precisely, that it does not take much to make an almost hereditary ring hereditary.

1.8 *Theorem.* The following are equivalent for a commutative ring A.

(1) A is hereditary

(2a) A is almost hereditary

(2b) The total ring of quotients K of A is hereditary.

Proof: That (1) implies (2) is clear. As in Ref.43, if I is an ideal of A, IK can be written $IK = \oplus\Sigma Ke_i$, e_i = idempotent. Since A is a domain at each localization A_M (M = prime ideal), the e_i's lie in A and thus $I \cap Ke_i = I \cap Ae_i = Ie_i$. On the other hand, $Ie_i \oplus A(1 - e_i)$ is an ideal of A clearly contained in no minimal prime ideal and is hence projective.

Since $I = \oplus \Sigma Ie_i$, we are through.

Remark. The closeness of hereditary rings to a finite direct product of Dedekind domains (*i.e.*, hereditary domains) reposes then on the behavior of its minimal primes. Thus, it will be Noetherian if such ideals are finite in number (since they are comaximal, the Chinese Remainder theorem takes over) or finitely generated [by Cohen's theorem and (1.6)]. There should exist, however, more natural conditions with the same consequence especially in the case $\text{Minimax}(A) = \emptyset$ (*i.e.*, no minimal prime is a maximal ideal). For although "hereditary" plus "$\text{Minimax}(A) = \emptyset$" does not necessarily imply that A is Noetherian (see examples in Ref.8), it comes very close since the nature of the minimal primes is largely determined by the maximal ideals in this case.

3. *Finitistic dimension one*

Although the rings with $FPD(A) = 1$ occur in various contexts, their general theory is rather sparse. After treating the decomposition of projective ideals into a finite product of projective primary ideals, we consider a condition (coherence) which often precipitates its Noetherianess.

1.9 *Theorem.* Let A be a commutative ring of finitistic projective dimension one. Then every projective ideal not contained in any minimal prime ideal is a product of projective primary ideals, in which the factors are unique up to order.

Proof: Let I be such ideal; by (1.6) it is a finitely generated faithful ideal. Thus according to the change of rings theorem in dimension one [35] for any A/I-module, E of finite projective dimension one has $pd_A E = 1 + pd_{A/I} E$ and thus $FPD(A/I) = 0$. Since A/I is commutative and perfect, we may write $A/I = B_1 \times \cdots \times B_n$, where B_i is a local ring of Krull dimension zero. This decomposition yields in A a representation

$$I = Q_1 \cdots Q_n$$

where Q_i is the primary ideal corresponding to the ith component of A/I. Note that each Q_i is finitely generated, by I plus $1 - e_i$, where e_i is some element of A mapping into the identity of B_i. As for the projectivity of Q_i, since it is a faithful ideal, it is enough [60] to show that it is principal at each localization A_M, which is clear. The uniqueness of the Q_i's is also immediate.

We recall the notion of a coherent ring. Let E be a finitely generated module over a ring A. E is said to be finitely presented if there is an exact sequence

$$A^m \longrightarrow A^n \longrightarrow E \longrightarrow 0$$

Such modules have extremely good properties with respect to localization and tensorization with infinite products. The rings for which every finitely generated ideal is finitely presented are called coherent.

1.10 *Proposition.* A ring is coherent if and only if the direct product of every family of flat modules is a flat module.

Proof: See Ref.15 for the proof and several other ways of expressing coherence.

1.11 *Proposition.* Let A be a commutative ring which is coherent and FPD(A) = 0. Then A is Artinian.

Proof: According to (0.19) we may assume that A is a local ring with a T-nilpotent maximal ideal P. To prove the statement, it is enough to show P finitely generated. From the coherence condition it suffices to show that P is the annihilator of a single element of A, which follows from the T-nilpotency of P.

The coherent rings of finitistic projective dimension one are not so lightly disposed of. In cases of domains however, we have the following proposition.

1.12 *Proposition.* Let A be a coherent domain with FPD(A) = 1. Then A is a Noetherian ring of Krull dimension one.

Proof: Let P be a nonzero prime ideal of A; pick $0 \neq a \in P$. The ring A/aA has, by the change of rings theorem of Ref.35, finitistic projective dimension zero. Since A/aA is also coherent, it follows from (1.11) that A/aA is Artinian. Thus P is finitely generated and A is Noetherian by Cohen's theorem.

4. *The minimal spectrum*

The rings of Tor-dimension one have the property of being domains at each localization, *i.e.*, they are rings in which principal ideals are flat (PF-rings for brevity). They will be PP-rings (for "principal ideals are projective" as in the case of rings of global dimension one) when some additional conditions are imposed on the topology of Min(A), the subspace of Spec(A) consisting of the minimal primes.

1.13 *Proposition.* The following conditions are equivalent for a commutative ring A:

(1) A is a PP-ring

(2a) A is a PF-ring

(2b) Min(A) is compact.

Proof: It will be enough to show that (2) implies (1), the reverse implication being clear. Let $f \in A$, and write I = annihilator of f. Since A is locally a domain, I is a pure ideal of A. On the other hand, the ideal $J = (f,I)$ is not contained in any minimal prime. Thus we can write

$$\mathrm{Min}(A) = \bigcup_{g \in J} D_g \qquad \text{with } D_g = \{P \in \mathrm{Min}(A),\ g \notin P\}$$

By the compactness of Min(A) we can write Min(A) as the union of finitely many D_g's, and thus there is an ideal $(f,h_1, \ldots, h_n)$, $h_i \in I$, not contained in any minimal prime. From the purity of I, $(h) = Ih$ for every $h \in I$. It is clear then that there exists $g \in I$ such that $h_i(1 - g) = 0$ for each h_i. In particular $L = (f,g)$ is not contained in any minimal prime. However, L is locally isomorphic to A and is thus projective. Since $(f) \cap (g) = 0$, $L = (f) \oplus (g)$ and (f) is also projective.

1.14 *Remarks.*

(a) For an example of a PF-ring with Min(A) not compact, see Example 1.3b. If T is an indeterminate over a ring A, Min(A) is homeomorphic to Min(A)[T]). Thus if A is a PP-ring, so is A[T].

(b) Still another way of expressing (1) or (2) above is "the total ring of quotients of A is a von Neumann ring."

(c) Later on we shall see that the compactness (or lack of it) of the minimal spectrum of a ring of global dimension two is a very important controlling factor in the choice of methods needed for a description of these rings.

(d) For the PF-rings of finitistic projective dimension one, Min(A) is compact. Indeed, if $f \in A$, I = annihilator of f is a pure ideal and $L = (f,I) = (f) \oplus I$ is locally principal - hence flat. Since $\mathrm{FPD}(A) = 1$, L must be projective; thus L is finitely generated and I is finitely generated also. For such rings we also have (i) Krull $\dim(A) \leq 1$, and (ii) the total ring of quotients of A has global dimension at most one.

5. *Generating ideals efficiently*

Here we give an estimate for the number of generators needed for finitely generated ideals in rings of global dimension or finitistic projective dimension one. Since Noetherian conditions on the spectrum of the ring are usually missing, we must proceed rather differently from the classical treatment [59]. We need the following lemma [22].

1.15 *Lemma.* Let I be a finitely generated ideal of a commutative ring A. If I/I^2 is generated by n elements, then I can be generated by n + 1 elements.

Proof: Let $a_1, \ldots, a_n$ be elements of I that generate I modulo I^2; denote the ideal the a_i's generate by J. In A/J the ideal L = I/L is such that $L^2 = (I^2 + J)/J = I/J = L$. Since L is finitely generated, it is generated by one idempotent. Thus I can be effectively generated by n + 1 elements.

1.16 *Theorem.* A finitely generated flat ideal of a ring A can be generated by two elements in the following cases:

(1) A has Krull dimension one.

(2) A has finitistic projective dimension one.

Proof: Let I be a finitely generated flat ideal, and write L for its annihilator. Since I is, at each localization, a free module, L is a pure ideal (recall that A/L is A-flat). From the exact sequence

$$0 \longrightarrow I \longrightarrow A \longrightarrow A/I \longrightarrow 0$$

tensoring with A/L, we conclude that

$$I/LI \hookrightarrow A/L \quad \text{or} \quad I \hookrightarrow A/L$$

Thus I may be viewed as a finitely generated ideal of A/L. The annihilator of I is now trivial, and consequently I is a faithful projective ideal of A/L [60]. Also note that A/L inherits the property (1) or (2) above.

According to (1.15), to show that I can be generated by two elements, we examine $E = I/I^2$ $(= I \otimes_A A/I)$.

Case (1): E is a rank one projective module over the ring B = A/I of Krull dimension zero (since I is faithful, it cannot be contained in any minimal prime ideal). Let J be the nilradical of B. E/JE is a rank one projective module over the von Neumann ring B/J. According to Ref.34 such modules are direct sums of principal ideals. Rank consideration then shows that E/JE is actually B/J-free. By Nakayama's lemma, E is principal and (1.15) applies.

Case (2): From the change of ring theorem of Ref.35, A/I is a ring of finitistic projective dimension zero and E is again principal.

1.17 *Corollary*. Every finitely generated ideal of a Prüfer domain of Krull dimension one can be generated by two elements.

Remarks. (1.17) is the one-dimensional case of a conjecture stating that every finitely generated ideal of a Prüfer domain can be generated by two elements. In Ref.28 this has been extended to show that finitely generated ideals in Prüfer domains of Krull dimension n can be generated by n + 1 elements. The arguments, however, are quite different.

A question that remains is whether in a ring of FPD(A) = 1 every rank one projective module (not just ideals) can be generated by two elements. If this were the case, the device introduced here could be used to estimate the number of generators needed for a projective ideal in a ring with FPD(A) = 2. In the Noetherian case this always works since a rank one projective module is isomorphic to an ideal. Using (0.16), this allows for the well-known estimate that a projective ideal of a Noetherian ring of Krull dimension n can be generated by n + 1 elements.

Chapter 2

LOCAL RINGS (I)

In this chapter we analyze the conditions that a local ring must satisfy if it is to have global dimension two. In Chapter 4 we use this analysis to construct rings of global dimension two.

1. *Classification*

We have already mentioned the class of Noetherian local rings of finite global dimension (see Theorem 0.15). Another family of local rings of finite global dimension is provided by the following theorem [46].

2.1 *Theorem.* Let V be a valuation domain; V has global dimension $n + 2$ if and only if every ideal of V can be generated by χ_n elements, and some ideal does need that many generators.

The main result here says that regular local rings of dimension two and valuation domains with countably generated ideals are the basic ingredients of an arbitrary local ring of global dimension two [63].

2.2 *Theorem.* A local ring of global dimension two is either a regular local ring, a valuation domain, or a so-called umbrella ring. A local ring A is an umbrella ring if A is a domain and contains a prime ideal P such that

(a) $P = PA_P$, that is, P is divisible by any outside element

(b) A/P is a two-dimensional regular local ring

(c) A_P is a valuation domain with countably generated ideals

(d) A has only countably many principal prime ideals.

Proof: The proof consists of a series of observations on A.

(1) A is a domain: In fact, if f is a nonzero element of A with annihilator I, the exact sequence

$$0 \longrightarrow I \longrightarrow A \longrightarrow Af \longrightarrow 0$$

says that I is A-projective and thus, by Theorem 1.4, free; but this is only possible if $I = 0$, since f annihilates I.

(2) A is a coherent ring: If I is a finitely generated ideal of A, there is an exact sequence

$$0 \longrightarrow K \longrightarrow A^n \longrightarrow I \longrightarrow 0$$

with K a free module and thus of rank necessarily not greater than n.

(3) A is a GCD domain; in particular, A is integrally closed: If a, b are nonzero elements of A, consider the sequence

$$0 \longrightarrow K \longrightarrow A^2 \longrightarrow (a,b) \longrightarrow 0$$

obtained by mapping (1,0) onto a and (0,1) onto -b. K will be a free module of rank one, generated by, say, (β,α). As $(b,a) \in K$, we can write $(b,a) = \delta(\beta,\alpha)$. It is easily checked that δ is the greatest common divisor of a and b.

We denote the GCD of a and b by [a,b]. Note that if $[a,b] = \delta$, $a = \delta\alpha$, $b = \delta\beta$, then $\{\alpha,\beta\}$ form a regular sequence, that is, $r\alpha = s\beta$ yields $r \in (\beta)$, $s \in (\alpha)$. Let M denote the maximal ideal of A.

(4) If M is generated by one element, A is a valuation domain: Assume $M = (d)$; if a, b are elements of M, $[a,b] = \delta$, then with $a = \delta\alpha$, $b = \delta\beta$, we must show that α or β is a unit. If this were not so, α and β would both be divisible by d and this is not possible by the remark above.

Remark. In Chapter 4 we shall see how steps 1 to 4 hold in any coherent local ring of finite Tor-dimension.

(5) If M is not finitely generated A is a valuation domain: Let a, b be elements of M. With the notation of (4) it is enough to show that the ideal $I = (\alpha,\beta)$ is the unit ideal. Assume $I \subset M$; notice that by the change of rings of Ref. 35 applied twice, as $\{\alpha,\beta\}$ is a regular sequence, the finitistic projective dimension of A/I is zero. From (2), A/I is also a coherent ring. Thus A/I by (1.11), is an Artinian ring and its maximal ideal, M/I, is finitely generated. Hence M is finitely generated.

Assume from this point on that M is a finitely generated ideal but not principal.

Before we proceed with the next step, for easy reference, we quote from Ref. 12; see also Remark 6.5.

Let A be a commutative ring and let I be a finitely generated ideal containing a nonzero divisor and admitting a resolution

$$0 \longrightarrow A^{n-1} \xrightarrow{u} A^n \longrightarrow I \longrightarrow 0$$

2.3 *Theorem*. The ideal I can be written as $I = dD$, where d is a regular element and D is the ideal generated by the minors of order $n - 1$ of the matrix u.

Let us continue with the examination of the maximal ideal M.

(6) M is generated by two elements: If M is minimally generated by $x_1, \ldots, x_n$, since $\text{pd } M = 1$, there is an exact sequence

$$0 \longrightarrow A^{n-1} \xrightarrow{u} A^n \longrightarrow M \longrightarrow 0$$

By (2.3) above, $M = d(d_1,\ldots,d_n)$ where the d_i's are the $(n - 1)$-minors of u. Since all the entries of u lie in M, $M = M^{n-1}$. By Nakayama's lemma, then, $n = 2$.

(7) Every prime ideal $P \neq M$ is a directed union of principal ideals (and thus a flat module): Let a, b be elements of P and let $\delta = [a,b]$; we show that $\delta \in P$. If this were not so, with the notation used before, $(\alpha,\beta) \subset P$. But for the same reasons as in (5), $A/(\alpha,\beta)$ would then be an Artinian ring with $P/(\alpha,\beta) \subset M/(\alpha,\beta)$ distinct prime ideals, and thus we get a contradiction.

(8) Every non-finitely generated prime ideal is contained in each finitely generated prime ideal: We have seen enough thus far to conclude that the finitely generated prime ideals other than M are principal. Let P be a prime ideal $\neq$ M, let (d) be a principal prime, and assume $P \not\subset (d)$.

$A/(d)$ is a coherent domain of finitistic dimension one, and thus by (1.12), Noetherian. Consequently, (P,d) is a finitely generated ideal. From (7) we can assume $(P,d) = (p,d)$, $p \in P$. If p is not a generator for P, from (7) again it follows that we can find $q \in P$ with $p = rq$, $r \in M$. We can, on the other hand, write $q = tp + vd$; but then $q = trq + vd$ or $(1 - tr)q = vd$ and $q \in (d)$, since $1 - tr$ is invertible. But then $p \in (d)$.

(9) The non-finitely generated prime ideals of A are linearly ordered: Let P, Q be noncomparable, non-finitely generated prime ideals. Let $a \in Q \setminus P$, $b \in P \setminus Q$. Then $\delta = [a,b]$ is in neither P nor Q, and hence $P + Q$ contains the regular sequence $\{\alpha,\beta\}$ with the same meaning of before. Now choose $d \in M \setminus M^2$ (possible because M is finitely generated) and observe that (d) is a prime ideal since d is an indecomposable element in a GCD domain. Thus by (8), $P + Q \subset (d)$, which is not possible since $[\alpha,\beta] = 1$.

At this point we pause to look at the picture that emerges from the analysis so far. Assume that A is not a valuation domain. Let P be the union of all non-finitely generated prime ideals. Because these are linearly ordered, P is a prime ideal (non-finitely generated).

If $P = 0$, A would be Noetherian and thus a regular ring. If $P \neq 0$, since P is divisible by every prime ideal that contains it, we have

(a) $P = PA_P$.

Let N be the ring A/P; N is a Noetherian ring of Krull dimension at least two. Since its maximal ideal is generated by two elements:

(b) N is a two-dimensional regular local ring.

As for statement (c), we first need the following lemma.

2.4 *Lemma.* Let A be a commutative domain. Then A is a valuation domain if and only if it is a GCD domain and the prime ideals are linearly ordered.

Proof: A valuation domain satisfies trivially the GCD condition. A ring with the above properties is local; let a, b be elements in its maximal ideal, and let $\delta = [a,b]$. Then with $a = \delta\alpha$, $b = \delta\beta$, $\{\alpha,\beta\}$ is a regular sequence in M if neither α nor β is a unit. If that were the case, let P (resp. Q) be a prime ideal minimal over (α) [resp. (β)]. Say P Q; then there is $s \notin Q$ and an integer n such that $s\alpha^n = r\beta$. Since $\{\alpha^n,\beta\}$ also form a regular sequence, $s \in (\beta)$, a contradiction.

We are now ready to make the following statement.

(c) A_P is a valuation domain of global dimension one or two.

The homological statement being obvious, the remainder follows from (9), (2.3), and (2.5).

2.5 *Lemma.* Let A be a GCD domain. If B is a flat epimorphic image of A, then B is also a GCD domain.

Proof: Let a/b, $c/d \in B$, where we may assume $[a,b] = [c,d] = 1$. By the flatness of B [39], $(b):a = bA$ and $(d):c = dA$ imply that $(a/b, c/d)B = (a,c)B = (a,c)A \otimes_A B$. Now the exact sequence

$$0 \longrightarrow K \longrightarrow A^2 \longrightarrow (a,c) \longrightarrow 0$$

as in (3) (by the GCD property of A) yields that K is a free module of rank one. Tensoring it by B, we conclude that the relations of $(a/b, c/d)$ form a free module of rank one, which is equivalent to saying that B is a GCD domain.

Figuratively, the spectrum of a ring of global dimension two being described looks like

M

.... o o o o ....

P

.

.

o

o

.

.

0

(hence the name, "umbrella" ring).

We now proceed to the last statement (d), that in this case (that is, $P \neq 0$) A contains only countably many principal prime ideals. (Warning: This is not the case, however, for regular local rings of dimension two as the example, $A = C[x,y]_{(x,y)}$, C = complex numbers, shows.)

Let P(A) be the set (up to associates) of the prime elements of A, and let S be the multiplicative set they generate. We show that up to units S is a countable set. Choose a nonzero element $x \in P$ and let J be the ideal in A generated by x/s, $s \in S$. Notice that J is indeed an ideal since each s divides P. J is isomorphic to the A-module T of the field of quotients Q of A generated by $1/s$, $s \in S$. But the projective dimension of a submodule such as T has been calculated [48] and is given in terms of the cardinality of the set S modulo units. Since $\text{pd } T = \text{pd } J \leq 1$, the result mentioned says that P(A) must be countable.

In Chapter 4 we will see how the conditions of (2.2) are readily realized. As an illustration, we prove that the following ring [45] is not of global dimension two.

Let $F = K(a_i,b_i)$, $i = 1, 2, \ldots,$ be an infinite pure transcendental extension with generators a_i, b_i over the prime field of characteristic 2. Let $R = F[[x,y]]$ be the formal power series ring in the indeterminates x, y over F. Let g be the automorphism of R defined by

$$g(x) = x, \qquad g(y) = y$$

$$g(a_i) = a_i + y(a_{i+1}x + b_{i+1}y)$$

and

$$g(b_i) = b_i + x(a_{i+1}x + b_{i+1}y)$$

Note that $g^2 = 1$. Let A be the subring of R of elements left fixed by g. The maximal ideal of A is generated by two elements, but A is not Noetherian [45]. Since an umbrella ring has Krull dimension at least three, A is not any of the types described here.

Given this description, a few questions arise:

1. Considering that the injective modules over both Noetherian rings and valuation domains are reasonably well known, can one obtain a similar description of the injective modules over an umbrella ring?

2. Let A be a local ring of global dimension two, and let G be a finite group of automorphisms of A operating linearly and faithfully on M/M^2 (which is not the case of the example considered above). For what groups are the rings of fixed elements still of global dimension two?

Chapter 3

CONDUCTORS, CARTESIAN SQUARES, AND PROJECTIVITY

This chapter is devoted to a discussion of the role of the conductor of a ring extension vis-a-vis the descent of projectivity. The technical fact proved says that an injective homomorphism of commutative rings descends the projectivity if it does so modulo the conductor. The cheap version, that with Noetherian hypothesis, of the descent of projectivity by a finite homomorphism proved in the general case by Gruson [25] then follows easily. From our current point of view, its more rewarding applications lie in the comparison of the global dimension of the rings in a cartesian square with a "large conductor." In this chapter, as an illustration, we take up the construction of local rings of finite global dimension with zero divisors. In the next chapter it will be used again in the construction of umbrella rings.

1. Conductors and the descent of projectivity

The following theorem is the projective analog of Ref. 19, which could be used to abbreviate the proof. At no great cost, however, in face of the strong hypotheses we give a complete proof.

3.1 *Theorem.* Let $h: A \to B$ be an injective homomorphism of rings, I an ideal of A, and E an A-module. Then E is A-projective if and only if $B \otimes_A E$ is B-projective, and E/IE is A/I-projective in the following cases:

(1) I is also a B-ideal.

(2) I is nilpotent.

Proof: (1) (i) $\mathrm{Tor}_1^A(A/I,E) = 0$: We must show that the natural map $I \otimes_A E \longrightarrow E$ is injective. This follows from the commutative diagram

$$\begin{array}{ccc} I \otimes_A E & \longrightarrow & E \\ \downarrow & & \downarrow \\ I \otimes_B B \otimes_A E & \longrightarrow & B \otimes_B B \otimes_A E \end{array}$$

where the vertical map on the left is the natural identification, while the lower horizontal map is injective by the B-flatness of $B \otimes_A E$.

(ii) $\mathrm{Tor}_1^A(B,E) = 0$: Let

$$(*) \qquad 0 \longrightarrow G \xrightarrow{f} F \longrightarrow E \longrightarrow 0$$

be exact with F A-free. By tensoring it with B, we get (from now on unadorned tensor products are taken over A)

$$(**) \qquad 0 \longrightarrow \mathrm{Tor}_1^A(B,E) \longrightarrow B \otimes G \xrightarrow{1 \otimes f} B \otimes F \longrightarrow B \otimes E \longrightarrow 0$$

Since $B \otimes E$ is B-projective, this sequence splits piecemeal; by tensoring it with B/I over B, we get the exact sequence

$$0 \to B/I \otimes_B \mathrm{Tor}_1^A(B,E) \to B/I \otimes_B B \otimes G \to B/I \otimes_B B \otimes F \to B/I \otimes_B B \otimes E \to 0$$

which can also be written

$$0 \to \mathrm{Tor}_1^A(B,E)/I\mathrm{Tor}_1^A(B,E) \longrightarrow B/I \otimes G/IG \longrightarrow B/I \otimes F/IF \longrightarrow B/I \otimes E/IE \longrightarrow 0$$

From (i) it follows then that $\mathrm{Tor}_1^A(B,E) = I\mathrm{Tor}_1^A(B,E)$. Now let x be an element of $\mathrm{Tor}_1^A(B,E)$; we can write $x = \Sigma a_i x_i$ with $a_i \in I$, $x_i \in \mathrm{Tor}_1^A(B,E) \hookrightarrow B \otimes G$. Thus it follows that x lies in the image of G in $B \otimes G$. But in the diagram

$$\begin{array}{ccc} G & \longrightarrow & F \\ \downarrow & & \downarrow \\ B \otimes G & \longrightarrow & B \otimes F \end{array}$$

the right vertical map is injective since $h: A \to B$ is injective and F is A-free. We then have $x = 0$.

(iii) (*) splits: Let ϕ be a splitting for (**), that is, $\phi(1 \otimes f) = 1$. On the other hand, let ψ be a splitting of

$$0 \longrightarrow G/IG \xrightarrow{f'} F/IF \longrightarrow E/IE \longrightarrow 0$$

The projectivity of F yields then a map $\theta: F \to G$ such that $\theta f = 1 + g$, with $g: G \to IG$.

From the product

$$[1 - (1 \otimes \theta)(1 \otimes f)][\theta(1 \otimes f) - 1] = 0$$

one gets

$$[(1 \otimes g)\phi + (1 \otimes \theta)](1 \otimes f) = 1$$

If we restrict the map $(1 \otimes g)\phi + (1 \otimes \theta)$ to F, we get $[(1 \otimes g)\phi + \theta]f = 1$, where $(1 \otimes g)\phi + \theta$ is actually a map from F into G. This completes the proof of (1).

(2) Say $I^n = 0$, and let $I_i = A \cap BI^i$. Then $I_i^2 \subset I_{i+1}$ and $I_n = 0$. By passing to A/I_{n-1} ($\hookrightarrow B/BI_{n-1}$), we reduce the question, by induction, to the case $I^2 = 0$.

(i) $\mathrm{Tor}_1^A(A/I,E) = 0$: This can be read off the diagram

$$\begin{array}{ccc} I \otimes E & \longrightarrow & E \\ \| & & \\ I \otimes_{A/I} E/IE & & \\ \downarrow & & \downarrow \\ IB \otimes_{A/I} E/IE & & \\ \| & & \\ IB \otimes_B B \otimes E & \longrightarrow & B \otimes E \end{array}$$

where the left vertical map is injective by the A/I-flatness of E/IE, while the lower horizontal map is injective by the B-flatness of $B \otimes E$.

(ii) (*) splits: Tensor (*) with A/I and get $\phi\colon F \to G$ such that $\phi f = 1 + g$, $g\colon G \to IG$. Since $g^2 = 0$, $(1 - g)\phi$ provides the desired splitting map.

Remarks. (a) In (2) above it is enough that I be T-nilpotent, since it follows from Ref. 26 that E is flat and the argument in (2) yields a splitting. (b) If A is Artinian, with I the radical of A, we see that any injective homomorphism descends the projectivity. (c) With k a field and $A = k[x_i^2, x_i^3]$, $i = 1, 2, \ldots$, $B = k[x_i\text{'s}]$, the conductor I of B in A is such that $A/I = k$. Thus the inclusion $A \to B$ descends projectivity.

3.2 *Corollary*. Let A, B be commutative rings under the conditions of (1) above, and let E be an A-module. If $B \otimes E$ and E/IE are finitely generated modules over B and A/I respectively, then E is finitely generated over A (descent of finiteness).

Proof: Select n large enough such that there is a sequence

$$A^n \longrightarrow E \longrightarrow C \longrightarrow 0$$

such that $B \otimes C = 0$ and $C/IC = 0$. Then C is A-projective of trace ideal say, J. However, the trace of $B \otimes C$ is JB. Since A is a subring of B, $C = 0$.

3.3 *Theorem*. Let A be a Noetherian ring, $h\colon A \to B$ an injective homomorphism of rings, and B finitely generated as an A-module. Let E be an A-module; if B E is B-projective, then E is A-projective.

Proof: The proof follows a familiar path [19]. Let I be a largest ideal of A such that E/IE is not A/I-projective. If $I \neq \sqrt{I} = J$, E/JE is A/J-projective and $B/JB \otimes E$ is also B/JB-projective ($A/J \hookrightarrow B/JB$), and by (3.1) we get a contradiction. Thus we may assume that A is reduced and such that E/LE is A/L-projective for each ideal $L \neq 0$. We have the canonical inclusion $A \to A' = \Pi(A/P_i)$, P_i running through the minimal primes. Let $B' = \Pi(B/P_iB)$; then $A' \otimes E$ is A'-projective. However, the conductor of A'/A is nonzero, and thus by (3.1), E is A-projective.

Assume then A to be a domain. We may take $B = A[y]$ and also a domain. Let $ay^n + \cdots + b = 0$ be the least degree equation satisfied by y. Then with $z = ay$, $A \to B_0 = A[z] \to A[y]$, B_0 is a free extension of A, and hence it is enough to show that $B_0 \otimes E$ is B_0-projective. From the hypothesis on A it follows that for every ideal $M \neq 0$ of B_0, $M \cap A \neq 0$ and $B_0/M \otimes E$ is B_0/M-projective. Also, since the conductor of B/B_0 is nonzero, a final application of (3.1) yields the desired conclusion.

2. Cartesian squares (I)

A commutative square of ring homomorphisms

$$\begin{array}{ccc} A & \xrightarrow{i_1} & V \\ {\scriptstyle i_2}\downarrow & & \downarrow{\scriptstyle j_1} \\ N & \xrightarrow[j_2]{} & K \end{array}$$

is said to be <u>cartesian</u> (or a pull-back) if A is the product of N and V over K, that is, if the homomorphism $A \to V \times N = B$ induces an isomorphism of A onto the subring of B of pairs (x,y) with $j_1(x) = j_2(y)$; or if i_1 induces an isomorphism of $\ker i_2$ onto $\ker j_1$ and j_2 an injection of $\operatorname{coker} i_2$ into $\operatorname{coker} j_1$.

If one of the homomorphisms j_1 and j_2 is surjective, one encounters a situation which appears in a natural way in problems of descent. Assume j_1 to be surjective; then i_2 is also surjective and the square is cocartesian, that is, j_2 and i_1 permit the identification of $N \otimes_A V$ and K. Moreover, if $I = \ker i_2$, $i_1(I)$ is an ideal of V.

This case allowed Milnor [44] to describe the projective modules over A in terms of cartesian squares of projective modules over the other vertices of the square.

Instead of considering the descent in $A \to V$, it will be more convenient to consider the descent of projectivity associated with $A \to B = V \times N$. In this case we are in the situation of (3.1), where $\ker i_2$ is a conductor ideal, that is, a common ideal of A and B. In this section we consider an example where this ideal is very large. In the next chapter we will have another example but with more delicate properties.

Let V be a valuation domain with a non-principal maximal ideal M. Consider the cartesian square

$$\begin{array}{ccc} A & \longrightarrow & V \\ \downarrow & & \downarrow \\ V & \longrightarrow & V/M \end{array}$$

where the maps onto V/M are the natural ones. Then A is the subring of $B = V \times V$ of pairs (a,b) with $a - b \in M$. A is a local ring with zero-divisors and maximal ideal $J = M \times M$; notice that J is the conductor of B in A and $A/J = V/M$.

3.4 *Theorem.* $\text{gl dim}(V) \leq \text{gl dim}(A) \leq \text{gl dim}(V) + 1$.

Proof: First we remark that the need for a non-principal maximal ideal M arises because, if $M = (d)$, we would have

$$0 \longrightarrow (d,0)A \longrightarrow A \longrightarrow (0,d)A \longrightarrow 0$$

exact and then $\text{pd}_A(0,d)A = \infty$.

Assume that we have shown that

(1) $\text{flat dim}_A(B) \leq 1$

and let

$$0 \longrightarrow L \longrightarrow P_{n-1} \longrightarrow \cdots \longrightarrow P_0 \longrightarrow I \longrightarrow 0$$

be a resolution of the ideal I, where P_i is A-projective and $\text{gl dim}(V) = n$. To show that L is also projective, tensor the resolution with B to obtain a B-projective resolution of $B \otimes_A I$, and (3.1) applies.

For the verification of (1), it is enough to show that $\text{Tor}_1^A(I,B) = 0$ for every ideal I of A; we may even take I finitely generated, say,

$$I = ((a_1,b_1), \ldots, (a_n,b_n))$$

If v denotes the valuation associated to V, by consideration of $\min\{v(a_i)\}$ one easily concludes that

$$I = (a,0)A \oplus (0,b)A$$

for appropriate a, b. We may thus assume in verifying (1) that I is principal, generated by $(a,0)$. A one-step resolution of this ideal is

$$0 \longrightarrow (0,M) \longrightarrow A \longrightarrow (a,0)A \longrightarrow 0$$

Tensoring it with B, we get

$$0 \longrightarrow \mathrm{Tor}_1^A(I,B) \longrightarrow (0,M) \otimes B \longrightarrow B \longrightarrow I \otimes B \longrightarrow 0$$

Let x be an element of $\mathrm{Tor}_A^1(I,B)$; we may assume that

$$x = (0,b) \otimes (c,0)$$

Because M is not principal, we may write $(0,b) = (0,d)(0,e)$, $d,e \in M$. Thus

$$x = (0,d) \cdot (0,e) \otimes (c,0) = (0,d) \otimes (0,e) \cdot (c,0) = 0$$

3.5 *Remark.* Thus, if V is a valuation domain with countably generated ideals [46], that is, gl dim(V) = 2, with a non-principal maximal ideal M, the ring

$$A = V \times_{V/M} V$$

has global dimension ≤ 3. That it is not two follows from (2.2), since A is not an integral domain.

3.6 *Remark.* Using the flat analog of (3.1) [19], we find easily the Tor-dim(A) to be 2. Observe that A is not a coherent ring.

3.7 *Remark.* This presentation of A as a cartesian square is a simplification by B.Osofsky [49] of her example of a local ring of finite global dimension with zero divisors.

Chapter 4

LOCAL RINGS (II)

In Theorem 2.2 local rings of global dimension two or less were identified. Such a ring is either a regular (Noetherian) local ring, a valuation ring, or a so-called umbrella ring, that is, a ring satisfying conditions (a) to (d) of (2.2). The analysis of umbrella rings is completed in this chapter. In particular it will permit the conclusion that rings (local) of global dimension two are χ_0-Noetherian and give a straightforward procedure for the construction of examples. More generally an F-ring is defined to be a domain A, having a prime ideal P such that A_P is a valuation ring and $PA_P = P$. An umbrella ring is an F-ring, and our analysis of the umbrella ring is couched in terms of the more general procedure accordingly. The second objective of this chapter is the computation of the global dimension of an arbitrary F-ring, A, in terms of its associated rings A_P and A/P. In our treatment here we follow Ref. 23 closely.

1. *F-rings*

4.1 *Definition.* A domain A is an F-ring if A has a prime ideal $P \neq 0$ such that

(a) A_P is a valuation ring.

(b) $P = PA_P$.

Such an ideal is called an F-ideal.

In saying $P = PA_P$ one means that the extension of P in A_P is the same as the image of P under the canonical map taking A into A_P.

Assume throughout this section that A is an F-ring and P is an F-ideal. The proof of the following proposition is straightforward.

4.2 *Proposition.* (i) The prime ideals of A contained in P are linearly ordered by inclusion.

(ii) The family of F-ideals of A is linearly ordered by inclusion. If A is a GCD domain, this family has a unique maximal element.

(iii) If P is not a maximal ideal, then P is not finitely generated.

(iv) If $Q \neq 0$ is a prime ideal properly contained in P, then QA_P is not finitely generated as an ideal in A_P.

(v) The ideal P contains no prime elements unless it is maximal.

The following lemma [54] is useful in proving finiteness of flat ideals.

4.3 *Lemma*. Let V be a local ring with maximal ideal M, and let $I \neq 0$ be a flat ideal of V. If $I \neq MI$, then I is principal, and conversely.

Proof: Let a be an element in $M \setminus MI$. For any element c in I, the relation $ac + (-c)a = 0$ implies that we may write

$$c = \Sigma b_n \alpha_n \qquad a = \Sigma d_n \alpha_n \qquad (\alpha_n \in I)$$

and $ab_n = cd_n$. Since not all d_n's can lie in M, we conclude $c \in (a)$.

Since an F-ideal P is A-flat, we have the following obvious result.

4.4 *Lemma*. For a torsion-free A-module M, $\mathrm{Tor}_1^A(M, A/P) = 0$.

2. *Cardinality conditions*

If A is a local domain with quotient field K (qf A = K), according to Ref. 37, $\mathrm{pd}_A K = 1$ if and only if K is countably generated as an A-module. Thus, if A is a local UFD with $A \neq K$, then $\mathrm{pd}_A K = 1$ if and only if A has at most countably many principal prime ideals.

4.5 *Proposition*. If a local ring A of global dimension two is not Noetherian, then it has at most countably many principal prime ideals.

Proof: If A is a valuation ring, then it has at most one nonzero principal prime ideal; hence we can assume that A is an umbrella ring with maximal F-ideal P.

From each nonzero principal prime ideal, choose a generator and let S be the multiplicative set they generate. Fix $0 \neq z \in P$, and let J be the ideal of A generated by z/s, $s \in S$. The ideal J is isomorphic to L, the submodule of K generated by $1/s$, $s \in S$.

If A has more than countably many principal prime ideals, then one knowns that L is not countably generated. Using Ref. 48, we conclude that $\mathrm{pd}_A M = 2$, and so $\mathrm{pd}_A J = 2$. Since J is an ideal of A, this is impossible.

4.6 *Corollary*. If A is an umbrella ring with maximal F-ideal P, then

(a) A/P has countably many principal prime ideals.

(b) $\mathrm{pd}_{A/P}(A_P/P) = 1$.

Proof: (a) follows from (4.5), whereas (b) follows from (a), when we see that A/P is a UFD and qf(A/P) = A_P/P.

4.7 *Definition.* A ring is χ_0-Noetherian if every ideal can be generated by countably many elements.

4.8 *Theorem.* A local ring of global dimension two is χ_0-Noetherian.

Proof: We can restrict our attention to an umbrella ring A having P as its maximal F-ideal. Since every ideal containing P is finitely generated, we confine our attention to ideals contained in P. Let J be such an ideal.

Case (1) JA_P is not finitely generated over A_P: Let $a_1, a_2, \ldots$ be a countable set of elements in J that generate JA_P over A_P, and let S be as in (4.5). Since $JA_P = J$, a_i/s is in J for all s in S, and so the set $T = \bigcup_{s\in S} \bigcup_{i=1}^{\infty} \{a_i/s\}$ generates J.

Case (2) JA_P is finitely generated over A_P: Write $JA_P = tA_P$ for some $t \in J$ and let $T = \{t(a/b) \mid a,b \in S \text{ and } t(a/b) \in J\}$, where S is as above. For $z \in J$ let $z = t(x/s)$. If $x \in P$, then x/s is in A. If $x \notin P$, then $x = uy$ with $y \in S$ and u is a unit in A; so $z = u(t(y/s))$. In either case, z is in the A-span of T, and so J is countably generated.

We now turn our attention to the second objective of this section, namely (4.10).

4.9 *Proposition.* Let A be an F-ring with F-ideal P, and let M be an A-module. Then M is projective if and only if $M \otimes_A A/P$ and M_P are projective as A/P and A_P modules, respectively.

Proof: If $M \otimes A/P$ and M_P are A/P and A_P projective, then the module $M \otimes (A/P \times A_P)$ is projective over $A/P \times A_P$. Since the image of P in $A/P \times A_P$ is an ideal and $M \otimes A/P$ is A/P-projective, one can apply Theorem 3.1.

4.10 *Theorem.* Let A be a local F-ring with F-ideal P such that

(1) A_P is of global dimension one or two.

(2) A/P is a regular local ring of global dimension two.

(3) $\mathrm{pd}_{A/P}(A_P/P) = 1$.

Then A is an umbrella ring.

Proof: Since P is not finitely generated, $\mathrm{pd}_A P \geq 1$, and so the global dimension of A is at least two.

Let J be an ideal of A. For $0 \neq z \in P$ we have J and zJ isomorphic, and so we can assume $J \subset P$. Let

$$0 \longrightarrow L \longrightarrow F \longrightarrow J \longrightarrow 0$$

be an exact sequence with F free. Tensoring with A/P and A_P yields exact sequences of A/P and A_P modules, respectively [see (4.4)]:

$$0 \longrightarrow L \otimes A/P \longrightarrow F \otimes A/P \longrightarrow J \otimes A/P \longrightarrow 0$$

$$0 \longrightarrow L \otimes A_P \longrightarrow F \otimes A_P \longrightarrow J \otimes A_P \longrightarrow 0$$

If JA_P is not finitely generated over A_P, then by (4.3), $PJ = J$; so $L \otimes A/P$ is projective as an A/P-module. If JA_P is finitely generated over A_P, we write $JA_P = tA_P$ for some $t \in J$ and observe that J/PJ embeds in $JA_P/PJ = JA_P/PA_PJ = JA_P/JA_PP = tA_P/tA_PP \simeq A_P/P$. Consider the exact sequence of A/P-modules where C is the cokernel of the embedding

$$0 \longrightarrow J/PJ \longrightarrow A_P/P \longrightarrow C \longrightarrow 0$$

Since $\mathrm{pd}_{A/P}(A_P/P) = 1$ if $\mathrm{pd}_{A/P}(J/PJ) \geq 2$, then $\mathrm{pd}_{A/P}C \geq 3$ [35]. Thus $\mathrm{pd}_{A/P}(J/PJ) \leq 1$, and so $L \otimes A/P$ is projective as an A/P-module.

In each case $L \otimes A/P$ is A/P-projective; since L_P is A_P-projective, L is projective and $\mathrm{pd}_A J \leq 1$. Thus A has global dimension two.

The following result is proved in a broader context in Section 4. It seems fitting, however, to record it here.

4.11 *Proposition*. Let A be a local F-ring with F-iceal P such that

(1) A_P is of global dimension one or two.

(2) A/P is a regular local ring of global dimension two.

(3) $\mathrm{pd}_{A/P}(A_P/P) = 2$.

Then A has global dimension three.

4.12 *Example*. Let R be a domain with quotient field Q. By R[[t)) one means the subring of Q[[t]] consisting of those power series whose constant term lies in R. Let P be the ideal of R[[t)) consisting of those elements whose constant term is zero. P is a prime ideal, $R[[t))_P = Q[[t]]$, a discrete valuation ring, and $PR[[t))_P = P$; hence R[[t)) is an F-ring with F-ideal P.

Let k be a field and let $R = k[x,y]_{(x,y)}$ so that $Q = k(x,y)$, and let A = R[[t)) with P as above. Thus, A_P is a valuation ring of global dimension one, and A/P is a regular local ring of global dimension two. If k is infinite, the cardinality of the set of principal prime ideals of R is the same as that of k. Thus, if k is countable, then gl dim(A) = 2; otherwise gl dim(A) = 3.

3. *Fiber products*

The aim of this section is to investigate a construction that yields F-rings and one from which every F-ring arises. We show that A is an F-ring with F-ideal P if and only if A is the fiber product of A_P and A/P over A_P/P.

4.13 *Construction.* Let V be a valuation ring with maximal ideal P, and let D be a subring of V/P which is not a field. Let A be the inverse image of D under the canonical map, π, taking V onto V/P. The diagram representing this situation is

$$(*) \qquad \begin{array}{ccc} A & \xrightarrow{\ i\ } & V \\ \pi \downarrow & & \downarrow \pi \\ D & \xrightarrow[\ j\ ]{} & V/M \end{array}$$

where i and j are the inclusion maps as indicated. One has that P is an ideal of A.

4.14 *Proposition.* Adopting the earlier notation, qf D = V/M if and only if $V = A_P$.

Proof: (if) Choose $u \in V/P$, and let $v \in V$ be such that $\pi(v) = u$. Write $v = a/b$ for $a,b \in A$ and $b \notin P$; hence $\pi(a/b) = \pi(a)/\pi(b)$. But $\pi(a),\pi(b) \in D$ since $a,b \in A$ and $\pi(b) \neq 0$. Thus $u = \pi(v) = \pi(a)/\pi(b)$, an element in the quotient field of D.

(only if) For $v \in V$ there exist $a_1,a_2 \in A$ with $a_2 \notin P$ such that $\pi(v) = \pi(a_1)/\pi(a_2)$. Since $\pi(v) = \pi(a_1/a_2)$, then $v - a_1/a_2 = z \in P \subset A$. Thus, $v = (a_2z + a_1)/a_2 \in A_P$.

4.15 *Basic construction.* If qf D = V/M in (4.13) we say we have the basic construction. We then say that (*) is a basic diagram and that A arises from the basic construction. If A arises from a basic construction then $V = A_P$, $D = A/P$, $V/P = A_P/P$, and $P = PA_P$.

4.16 *Proposition.* A ring A is an F-ring if and only if it arises from a basic construction.

Proof: Immediate.

4.17 *Remark.* The construction (4.13) is usually discussed without the hypothesis that qf D = V/P. The rings considered here, however, need not be of the form D + M as illustrated by (4.20).

We now restate (4.15) to conform to the situation considered by Milnor [44].

4.18 *Proposition.* Let

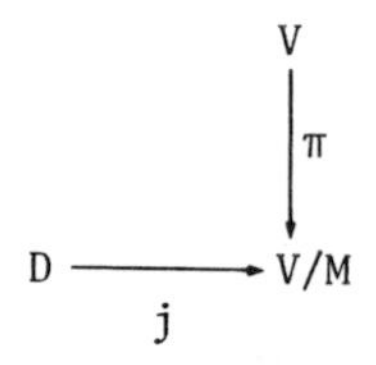

be a basic diagram. The pull-back of j and π is equal to the ring arising from the basic construction.

4.19 *Corollary.* A local ring A is an umbrella ring if and only if it is the pull-back of a basic diagram in which

(1) V has global dimension one or two.

(2) D is a regular local ring of global dimension two.

(3) $pd_D\ V/P = 1$.

We now have the characterization of local rings of global dimension two. A local ring of global dimension two can be a regular local ring or a valuation domain; if neither of these, it is the fiber product of them.

4.20 *Example.* Let $\mathbb{Z}$ denote the integers, and let q be a prime number. Then $Q = q\mathbb{Z}[x,y]$ is a prime ideal in $\mathbb{Z}[x,y]$ and the ring $V = \mathbb{Z}[x,y]_Q$ is a DVR with maximal ideal denoted by P. We now define $D = \mathbb{Z}/q\mathbb{Z}[x,y]_{(x,y)}$; the quotient field of D is isomorphic to V/P.

The fiber product of V and D over V/P as in (4.13) is

$$A = \{f/g + q(F/G) \mid q \nmid G \text{ and } q \nmid g(0,0), \text{ where } f,g,F,G \in \mathbb{Z}[x,y]\}$$

A has global dimension two [see (4.19)] and is of unequal characteristic and so cannot be expressed in the form $D + M$.

4. *Global dimension of F-rings*

The aim of this section is to find the global dimension of an F-ring A having F-ideal P in terms of the global dimensions of the rings A/P and A_P. We conclude with a few cardinality observations. Throughout this section A will be an F-ring and P an F-ideal.

4.21 *Proposition.* If M is a torsion-free A-module, then $pd_A M = \max\{pd_{A/P} M \otimes A/P,\ pd_{A_P} M \otimes A_P\}$.

Proof: Let $m = \max\{pd_{A/P} M \otimes A/P,\ pd_{A_P} M \otimes A_P\}$. If $pd_A M$ is finite, tensoring a projective resolution for M by A_P and A/P yields projective resolutions for $M \otimes A_P$ and $M \otimes A/P$ as A_P and A/P modules, respectively, and so $m \leq pd_A M$.

If m is finite, consider the exact sequence of A-modules

$$0 \longrightarrow Q \longrightarrow Q_{m-1} \longrightarrow \cdots \longrightarrow Q_0 \longrightarrow M \longrightarrow 0$$

where Q_i is projective for i = 0, 1, ..., m-1. But $Q \otimes A_P$ and $Q \otimes A/P$ are both projective as A_P and A/P modules, respectively, and so $pd_A M \leq m$.

If either $pd_A M$ or m is infinite, so is the other.

4.22 *Corollary*. $pd_A A_P = pd_{A/P}(A_P/P)$.

4.23 *Theorem*. If

(1) A_P has global dimension n, $n \geq 1$

(2) A/P has global dimension m, $m \geq 1$

$$\text{A has global dimension} \begin{cases} n & \text{if } n > m \\ m & \text{if } m \geq n \text{ and } pd_{A/P}(A_P/P) < m \\ m+1 & \text{if } m \geq n \text{ and } pd_{A/P}(A_P/P) = m \end{cases}$$

Proof: (i) $n > m$. For I an ideal in A, $pd_A I \leq n - 1$, and so gl dim(A) $\leq n$. There exists an ideal J in A such that $pd_{A_P} JA_P = n - 1$, and so $pd_A J = n - 1$; hence gl dim(A) = n.

(ii) $m \geq n$. Let I be an ideal in A (containing P) such that $pd_{A/P} I/P = m - 1$. Since $I/P = I/IP$, $pd_A I = m - 1$, and so gl dim(A) $\geq m$.

Let I be an arbitrary ideal in A. Multiplying I by a nonzero element in P allows us to assume I is contained in P in computing its projective dimension. Consider the exact sequence where F is free:

$$0 \longrightarrow L \longrightarrow F \longrightarrow I \longrightarrow 0$$

Tensoring with A_P and A/P yields exact sequences of A_P and A/P modules, respectively. One sees $pd_{A_P} I \otimes A_P \leq n - 1 \leq m - 1$.

If IA_P is not finitely generated over A_P, mimicking (4.10) and using (4.22) yields that $pd_A I \leq m - 1$.

If IA_P is finitely generated over A_P, as in (4.10) one has the exact sequence

$$0 \longrightarrow I/PI \longrightarrow A_P/P \longrightarrow C \longrightarrow 0$$

(a) If $pd_{A/P}(A_P/P) < m$, then $pd_{A/P} I/PI \leq m$; otherwise $pd_{A/P} C \geq m + 1$, and thus $pd_A I \leq m - 1$. Therefore gl dim(A) = m.

(b) If $pd_{A/P}(A_P/P) = m$, then $pd_{A/P} I/PI \leq m$; otherwise $pd_{A/P} C \geq m + 2$, and thus $pd_A I \leq m$. Therefore gl dim(A) $\leq m + 1$. But $pd_A A_P = m$,

(4.22). Choosing $0 \neq z \in P$, zA_P is an ideal of A having projective dimension m, and thus gl dim(A) = m + 1.

The next two corollaries follow directly from (4.23).

4.24 *Corollary.* If gl dim(A) = d, then

(1) gl $\dim(A_P) \leq d$.

(2) gl $\dim(A/P) \leq d$.

(3) If gl dim(A/P) = d, then $pd_{A/P}(A_P/P) < d$.

4.25 *Corollary.* Let gl dim(A) = d.

(1) If gl $\dim(A_P) = d$, then gl $\dim(A/P) \leq d$; and if gl dim(A/P) = d, then $pd_{A/P}(A_P/P) < d$.

(2) If gl $\dim(A_P) < d$, then

(a) gl dim(A/P) = d if $pd_{A/P}(A_P/P) < d - 1$

(b) gl dim(A/P) = d - 1 or d if $pd_{A/P}(A_P/P) = d - 1$.

Let us now consider a few cardinality conditions for an F-ring of global dimension two.

4.26 *Remark.* An F-ring of global dimension two need not have only countably many principal prime ideals even though it will have only that many locally. For, if M is a maximal ideal of A, then A_M is a local F-ring of global dimension at most two, and so has only countably many principal prime ideals.

4.27 *Example* (Jensen). Let C denote the complex numbers and let A = C[x][[t)). If we let V = C(x)[[t]], then V is a valuation domain of global dimension one with maximal ideal P = (t). Let D = C[x] so D is contained in V/P and qf D = V/P. The ring A is the pull-back as usual and so has global dimension two; yet it has uncountably many principal prime ideals.

In this example gl dim(A/P) = 1. Continuing as in (4.23), suppose we now have gl dim(A/P) = 2 and A/P is regular. Then locally A/P has only countably many principal prime ideals since $pd_{A/P}(A_P/P) = 1$. Is it the case that A/P has only countably many invertible prime ideals?

4.28 *Question.* Does there exist a regular ring of global dimension two having uncountably many invertible prime ideals yet having only countably many at each localization at a prime ideal?

The answer to this question will determine whether A need be χ_0-Noetherian under the assumptions above.

4.29 *Remark*. An alternate approach to the proof of (4.9) based on Ref. 44, Theorem 2.2, is provided here enabling us to bypass (3.1). We depend however upon the fact that A is the fiber product of A_P and A/P over A_P/P.

Adopting the notation of (4.9), we assume in addition that M is torsion free. Tensoring the cartesian square that defines A by M yields a cartesian square for M since $Tor_1^A(A/P,M) = 0$:

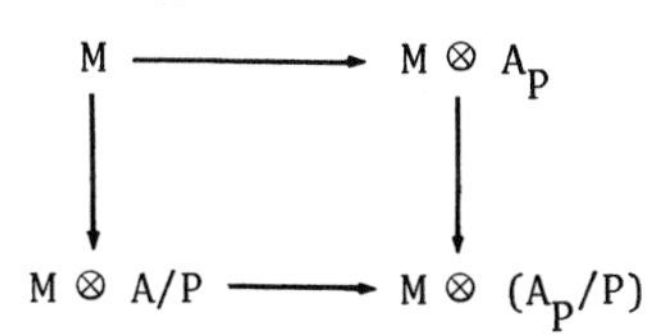

According to Ref. 44, the projectivity of $M \otimes A_P$ and $M \otimes A/P$ as A_P and A/P modules, respectively, implies the projectivity of M.

Chapter 5

CHANGE OF RINGS AND DIMENSIONS

Given a ring homomorphism $h\colon A \longrightarrow B$ and a B-module E, in this chapter we compare the projective dimensions (or some other homological dimension) of E when view both as a B-module and as an A-module via h. For most applications, B is a quotient A/I.

A first relation between $\mathrm{pd}_A E$ and $\mathrm{pd}_B E$ is

$$\mathrm{pd}_A E \leq \mathrm{pd}_A B + \mathrm{pd}_B E \tag{1}$$

which results from the spectral sequence [14]

$$E_2^{p,q} = \mathrm{Ext}_B^p(E,\mathrm{Ext}_A^q(B,C)) \underset{p}{\Longrightarrow} \mathrm{Ext}_A^n(E,C) \tag{2}$$

where C may be any A-module. If $d = \mathrm{pd}_A B$ and $e = \mathrm{pd}_B E$ are both finite, there are several instances where this spectral sequence allows the conclusion $E_2^{d,e} \neq 0$, and hence equality in (1).

Here we will study the case where B is not only finitely generated as an A-module but admits a resolution

$$0 \longrightarrow P_d \longrightarrow \cdots \longrightarrow P_1 \longrightarrow P_0 \longrightarrow B \longrightarrow 0 \tag{3}$$

where the P_i's are finitely generated projective modules.

In this case the module $\mathrm{Ext}_A^d(B,A)$ plays a significant role, and we begin with an elementary discussion of it. A first property to notice is that if C is any A-module, we have a canonical isomorphism

$$\mathrm{Ext}_A^d(B,C) = \mathrm{Ext}_A^d(B,A) \otimes_A C$$

This arises either from a spectral sequence argument or more simply from the following considerations. Since only the module structure of B is involved, we may assume that $\mathrm{pd}_A B = 1$. Write then a resolution

$$0 \longrightarrow P_1 \xrightarrow{\ f\ } P_0 \longrightarrow B \longrightarrow 0$$

with P_0, P_1 finitely generated projective modules. Applying $\mathrm{Hom}_A(-,C)$, we get the exact sequence

$$\mathrm{Hom}_A(P_0,C) \longrightarrow \mathrm{Hom}_A(P_1,C) \longrightarrow \mathrm{Ext}^1_A(B,C) \longrightarrow 0$$

The result follows from observing the natural equivalence of $\mathrm{Hom}_A(P_i,-) = \mathrm{Hom}_A(P_i,A) \otimes_A (-)$, since P_i is projective and finitely generated.

We also consider two types of conditions on B; the module $\mathrm{Ext}^d_A(B,A) = T$ plays an essential role for both of them, and they are, under certain circumstances, necessary for equality in (1). We shall abbreviate this situation at times, by saying that there is change of rings.

1. Macaulay extensions

We assume here that

$$\mathrm{Ext}^i_A(B,A) = 0 \qquad \text{for } i < d \qquad \text{(Condition M)} \quad (4)$$

This restriction is reminiscent of the condition a factor ring of a regular local ring must satisfy to be a Macaulay ring [36].

In this case, if we apply $\mathrm{Hom}_A(-,A)$ to the sequence (3) twice, taking into account the reflexivity of the projective modules, and the maps between them, we obtain the isomorphism of B-modules $\mathrm{Ext}^d_A(T,A) \simeq B$. In particular, we conclude that T is a faithful B-module.

5.1 *Theorem.* Let B be an extension of A satisfying conditions (3) and (4) above. Then for any B-module E with $\mathrm{pd}_B E < \infty$, we have

$$\mathrm{pd}_A E = \mathrm{pd}_A B + \mathrm{pd}_B E$$

Proof: According to (2) it will be enough to show that for some module $C = \oplus\, \Sigma A = A^{(I)}$,

$$\mathrm{Ext}^e_B(E,\mathrm{Ext}^d_A(B,C)) = \mathrm{Ext}^e_B(E,\, T \otimes_A C) = \mathrm{Ext}^e_B(E,T^{(I)}) \neq 0$$

This will be accomplished by an argument due to Gruson [25].

5.2 *Proposition.* Let E be a finitely generated faithful module over the commutative ring A. Then every A-module has a finite composition series whose factors are quotients of direct sums of copies of E. Thus every such module is piecewise a generator for the category of A-modules.

Proof: It is enough to show this for A itself: Let $I_0 \subset I_1 \subset \cdots \subset A$ be a sequence of ideals of A such that for each i there is a surjection

$$\phi_i\colon E^{(\Gamma_i)} \longrightarrow I_i/I_{i-1} \longrightarrow 0$$

Let $\phi\colon A^{(\Gamma)} \longrightarrow M$ be a presentation for a module M. Then the series of submodules of M given by $\phi^{-1}(I_i^{(\Gamma)})$ has the desired properties.

For the proof in the ring case, we recall the definition of the

Fitting invariants of the module E [36]: Let

$$A^{(I)} \xrightarrow{u} A^n \longrightarrow E \longrightarrow 0$$

be a free presentation of E and denote by $(u_{ij})_{1 \leq j \leq n,\ i \in I}$ the matrix of u. Let $F_r = F_r(E)$ be the ideal generated by the minors of order $n - r$ of this matrix. We note the following properties of these ideals [36]:

(1) $F_0 \subset F_1 \subset \cdots \subset F_n \subset F_{n+1} = A$

(2) If J is the annihilator of E, $J^n \subset F_0 \subset J$.

The smallest integer r, such that $F_r \neq 0$, exists and is nonzero; the smallest integer m, such that $F_m = A$, exists and $m \geq r$. We note that $m = r$ means that E is a projective module [60], and in this case we are done.

We shall reason by induction on $d = m - r$.

Suppose $d > 0$. Let $I = \mathrm{Ann}_A(E/F_rE)$. We can apply the induction hypothesis to the A/I-module E/IE: (a) this module is faithful $[xE \subset IE \subset F_rE \Rightarrow x \in I]$, and (b) its rth Fitting ideal is trivial and its mth Fitting ideal is A/I. On the other hand, by property (2) above, $I^n \subset F_r$. This gives rise to the filtration

$$F_r \subset F_r + I^{n-1} \subset \cdots \subset I$$

where each factor is an A/I-module. Thus to complete the proof, it is enough to show that F_r is a quotient of a direct sum of copies of E.

Let D be the determinant of the $n - r = s$ minor defined by $(u_{i_k, j_\ell})_{1 \leq k,\ell \leq s}$, and let us look for a linear form on E whose image contains D. One chooses an integer t between 1 and n and distinct from the i_k's: this is possible since $r > 0$. We consider the following linear form on A^n:

$$(x_1, \ldots, x_n) \longrightarrow \det \begin{vmatrix} a_{i_1,j_1} & \cdots & a_{i_1,j_s} & x_1 \\ \cdot & \cdot & \cdot & \cdot \\ a_{i_s,j_1} & \cdots & a_{i_s,j_s} & x_s \\ a_{i_t,j_1} & \cdots & a_{i_t,j_s} & x_t \end{vmatrix}$$

Since the minors of order $s + 1$ are zero, this form is trivial on the image of u; it defines then, by passage to the quotient, a linear form on E. On the other hand, it transforms the tth vector of the canonical basis of A^n into D.

We point out some immediate consequences.

5.3 *Corollary*. Let F be a right exact functor which commutes with arbitrary direct sums and is trivial on E; then F is trivial.

In the situation of (5.1), consider the right exact functor on the category of B-modules,

$$F(-) = \mathrm{Ext}^d_B(E, T \otimes_B (-))$$

and use the result above to complete the proof.

5.4 *Remark*. For the injective and flat analogs of (5.1) one can proceed in a similar way to show the equality in the dimension formula using the following spectral sequences

$$E_2^{p,q} = \mathrm{Ext}^p_B(\mathrm{Tor}^A_q(B,C),E) \underset{p}{\Longrightarrow} \mathrm{Ext}^n_A(C,E)$$

and

$$E_2^{p,q} = \mathrm{Tor}^B_p(E,\mathrm{Tor}^A_q(B,C)) \underset{p}{\Longrightarrow} \mathrm{Tor}^A_n(C,E)$$

respectively.

Let us indicate how to proceed in the injective case (one uses the same type of argument, and module, in the flat case). Let M be an injective A-module. Because $\mathrm{Ext}^i_A(B,A) = 0$, $i < d$, it follows in much the same way as before that $\mathrm{Tor}^A_d(B,M) = \mathrm{Hom}_A(T,M)$. Let $\mathrm{Ext}^e_B(C,E) \neq 0$ for some B-module C; by (5.3), $\mathrm{Ext}^e_B(T,E) \neq 0$ also. Let

$$0 \longrightarrow T \longrightarrow M$$

be an A-injective envelope of T. $\mathrm{Ext}^{d+e}_A(M,E) = 0$ implies that the module $\mathrm{Ext}^e_B(\mathrm{Hom}_A(T,M),E) = 0$. Since T is finitely generated and faithful as a B-module, there is an exact sequence

$$0 \longrightarrow B \longrightarrow T^n$$

obtained by mapping 1 into $(x_1,\ldots,x_n)$ where the x_i's form a generating for T. But in this case, $\mathrm{Ext}^e_B(\mathrm{Hom}_A(B,M),E) = 0$ also and then the module $\mathrm{Ext}^e_B(\mathrm{Hom}_A(B,T),E) = 0$. Since there is a surjection

$$\mathrm{Hom}_A(B,T) \xrightarrow{\phi} T \longrightarrow 0$$

given by $\phi(f) = f(1)$, we contradict $\mathrm{Ext}^e_B(T,E) \neq 0$.

5.5 *Remark*. The theorem above applies in the following cases:

(1) $B = A/(x)$, x a nonzero divisor, to give Ref. 35, Theorem 3.

(2) $B = A/I$, where I is a faithfully projective ideal. According to (1.6) such ideals are finitely generated and we obtain the result of Ref. 31.

Theorem 0.16 puts very strict conditions on $B = A/I$ in order for change of rings to apply for this pair of rings. In particular, if A is a regular local ring, then change of rings applies to A/I if and only if it is a Macaulay ring, that is, if the condition (M) holds.

The particular case we shall need [64] concerns ideals of projective dimension one in rings of global dimension two.

5.6 *Definition.* We say that a finitely generated ideal I of a commutative ring A is overdense if the canonical map $A \longrightarrow \mathrm{Hom}_A(I,A)$ is an isomorphism.

Such ideals are faithful and if the ring A is Noetherian, this is equivalent to asking that a regular sequence of length two be inside I [5]. These ideals tend to abound in rings of global dimension two which also have Tor-dimension two.

5.7 *Theorem.* Let A be a commutative ring, and let I be an overdense ideal with $\mathrm{pd}_A I = 1$. If E is an A/I-module with $\mathrm{pd}_{A/I} E < \infty$, then

$$\mathrm{pd}_A E = 2 + \mathrm{pd}_{A/I} E$$

Proof: Let

$$0 \longrightarrow K \longrightarrow A^n \longrightarrow I \longrightarrow 0$$

be a projective resolution of I. We have to show that (a) K is finitely generated and (b) $\mathrm{Ext}^i_A(A/I,A) = 0$, $i = 0, 1$.

That (a) holds will follow from the fact that it is a projective module of constant rank $n - 1$, I being faithful.

As for (b), consider the sequence

$$0 \longrightarrow I \longrightarrow A \longrightarrow A/I \longrightarrow 0$$

that yields

$$0 \longrightarrow \mathrm{Hom}(A/I,A) \longrightarrow \mathrm{Hom}(A,A) \longrightarrow \mathrm{Hom}(I,A) \longrightarrow \mathrm{Ext}^1_A(A/I,A) \longrightarrow 0$$

As an application of the change of rings for Tor-dimension, consider the following example.

5.8 *Example.* Let P be a finitely generated prime ideal of $B = A[x]$, where A is a Prüfer domain and x is an indeterminate over A. If P is not an invertible ideal, then P is maximal. First notice that as B is coherent

[26] and Tor-dim(B) = 2 (from Theorem 0.14), $pd_B P = 1$. It is easy to check, or will follow from (6.5), that P is overdense. Thus by the change of rings, the finitistic flat dimension of B/P is 0. However, since this ring is a domain, P must be maximal.

2. *Finitely generated modules*

If E itself, in the previous considerations, is finitely generated, there is no need to impose such a strong restriction on B.

5.9 *Theorem.* Let A and B be rings, with B local and finitely generated as an A-module. Assume that B admits a finite resolution by finitely generated projective A-modules. Let E be a B-module admitting a finite free resolution over B. Then

$$pd_A E = pd_A B + pd_B E$$

This result [38], contrary to (5.1), has an elementary via of attack through the device of minimal projective resolutions.

Proof: Using our previous notation, notice that

$$\mathrm{Ext}_B^e(E,\mathrm{Ext}_A^d(B,A)) = \mathrm{Ext}_B^e(E,B) \otimes_B \mathrm{Ext}_A^d(B,A)$$

since E has a resolution by finitely generated projective modules. Since both modules in the product are nonzero and the ring B is local, the statement follows.

3. *Some equalities of dimensions*

A consequence of section 2 is that it provides estimates for the finitistic projective dimension of the ring B, to wit $FPD(B) \leq FPD(A) - pd_A B$.

Here, by restricting to the case of the fPD (that is, where one takes in the definition of FPD only those modules which have a resolution by finitely generated projective modules), we get a case of equality [52]. At the same time we shall initiate a discussion of the local rings of Tor-dimension two.

In Chapter 2, of considerable help was the fact that local rings of global dimension two were coherent domains. In Chapter 3 we saw an example of a ring of Tor-dimension two but not coherent. We shall see that coherence in local rings of Tor-dimension two implies not only its integrality but also the steps 1 to 4 of the proof of (2.2).

5.10 *Theorem*. Let A be a coherent local ring, and let I be a finitely generated ideal with $pd_A I < \infty$. Then

$$fPD(A) = pd_A(A/I) + fPD(A/I)$$

When A is Noetherian, there is a simple way (via the notion of depth [5]) to verify the equality. We need the following easy lemma. For the rest of this section we shall unspecified A-modules to be of finite presentation.

5.11 *Lemma*. Let {A,M} be a coherent local ring and let E be a finitely presented A-module. Then $pd_A E = n$ if and only if $Tor_n^A(E,A/M) \neq 0$, and $Tor_{n+1}^A(E,A/M) = 0$.

Proof: Observe that there is a resolution

$$\cdots\ F_2 \xrightarrow{f_2} F_1 \xrightarrow{f_1} F_0 \longrightarrow E \longrightarrow 0$$

where each F_i is a free module of finite rank and the matrix f_i has all of its entries in M. The result follows in just the same way as in the Noetherian case, say, dimension shifting and Nakayama's lemma.

Proof of (5.10): Since A/I is also a coherent ring, $fPD(A) \geq pd_A(A/I) + fPD(A/I)$ follows from (5.9).

Write $d = pd_A(A/I)$, let E be an A-module of finite projective dimension, and consider the minimal resolution

$$0 \longrightarrow N \longrightarrow F_{d-1} \longrightarrow \cdots \longrightarrow F_0 \longrightarrow E \longrightarrow 0$$

where several of the terms above could be trivial. Since $Tor_{i+d}^A(E,A/I) = Tor_i^A(N,A/I)$, $Tor_i^A(N,A/I) = 0$ for $i > 0$.

Consider the spectral sequence

$$E_{p,q}^2 = Tor_p^{A/I}(A/M, Tor_q^A(N,A/I)) \underset{p}{\Longrightarrow} Tor_n^A(A/M,N)$$

Since $E_{p,q}^2 = 0$, $q > 0$,

$$Tor_p^{A/I}(A/M,N/IN) \simeq Tor_p^A(A/M,N)$$

By (5.11) this isomorphism says that $pd_A N = pd_{A/I}(N/IN)$, and if $pd_A E > d$, we conclude that $pd_A E = d + pd_{A/I}(N/IN)$ to get the inequality $fPD(A) \leq d + fPD(A/I)$.

5.12 *Corollary*. Let {A,M} be a coherent local ring, and let I be a finitely generated ideal such that (1) $pd_A I < \infty$, and (2) the radical of I is M. Then $fPD(A) < \infty$.

Proof: According to (5.10) (by passing to the ring A/I), we may assume that A is a primary ring. Suppose, say, that

$$0 \longrightarrow F_1 \xrightarrow{f} F_0 \longrightarrow E \longrightarrow 0$$

is a minimal resolution of a module E. By McCoy's theorem [36] the minors of order $\text{rank}(F_1)$ of f have a nontrivial annihilator, which is impossible in our situation unless $F_1 = 0$.

Observe that (5.12) is valid under the condition that every proper finitely generated ideal of A/I has a nontrivial annihilator, again by McCoy's theorem. Those are the local rings in which the maximal ideal lies in the assassinator of A [39]. We shall use this remark in one case that involves a maximal ideal that is minimal over an ideal $(x):b$, with x a nonzero divisor. Then with $I = (x)$, it will follow that $\text{fPD}(A) = 1$.

5.13 *Remark.* Let A be a coherent ring. The determination of the Tor-dimension of A can be changed into the determination of that dimension of its localizations,

$$\text{Tor-dim}(A) = \sup\{\text{Tor-dim}(A_P)\}$$

One can, in fact, discard some of these primes. In $X = \text{Spec}(A)$, consider

$$Y = \{P \in X \mid \text{Spec}(A_P) \setminus \{P\} \text{ is quasi-compact in } \text{Spec}(A_P)\}$$

Claim: $\phi: A \longrightarrow B = \prod_{P \in Y} A_P$ is faithfully flat. Granted this, we obviously have

$$\text{Tor-dim}(A) = \sup\{\text{Tor-dim}(A_P),\ P \in Y\}$$

Proof of claim: B is flat by the coherence of A. Suppose $PB = B$ for some prime P of A. This means that there are $x_1, \ldots, x_n$ in P so that $(x_1, \ldots, x_n)$ is not contained in any prime of Y, clearly a contradiction.

A consequence of this remark, later to be improved upon, follows: Let A be a coherent ring with

$$\text{gl dim}(A) = \text{Tor-dim}(A) = 2$$

Then there is a maximal ideal $P \in Y$, because otherwise every A_P, $P \in Y$ would be a valuation domain [by (2.2)].

4. *Divisibility*

In order to study divisibility in a coherent local ring of finite Tor-dimension, we examine first the nature of the annihilator of a module with a finite resolution by finitely generated free modules.

Let E be such a module over the commutative ring A:

$$0 \longrightarrow F_n \longrightarrow \cdots \longrightarrow F_1 \xrightarrow{u} F_0 \longrightarrow E \longrightarrow 0$$

Define the Euler characteristic of E [62] as

$$\chi(E) = \Sigma(-1)^i \mathrm{rank}(F_i)$$

5.14 *Proposition.* The following statements hold:

(a) $\chi(E) \geq 0$.

(b) $\chi(E) = 0$ if and only if $\mathrm{Ann}(E) \neq 0$.

(c) $\chi(E) > 0$ implies that $\mathrm{Ann}(\mathrm{Ann}(E)) = 0$.

Proof: If S is a multiplicative set of A, clearly $\chi(E_S) = \chi(E)$. If we then take $S = A \setminus P$, with P a minimal prime ideal, then E_P is a module with a finite free resolution over a ring in which every proper finitely generated ideal has a nontrivial annihilator. (Let us call a ring with this property a 0-ring). It follows from Ref. 39 that E_P is free. Thus $\chi(E_P) \geq 0$, and (a) is proved.

Let J be the 0th Fitting invariant of E, that is, J is the ideal (finitely generated) generated by the minors of order $r = \mathrm{rank}(F_0)$ of the matrix of u. Recall that if I is the annihilator of E,

$$I^r \subset J \subset I$$

If $I \neq 0$, let P be minimal over L, the annihilator of a nonzero element x of I. Localizing at P we get A_P, a 0-ring. As remarked earlier, E_P is then a free module over A_P. Since $0 \neq x \in A_P$, we get $E_P = 0$. Thus $\chi(E) > 0$ forces $I = 0$. Conversely, by localizing at a minimal prime ideal we conclude the converse part of (b)

Suppose now $\chi(E) = 0$, that is, $I \neq 0$, and let x be a nonzero element of Ann(I). In this case, let P be a prime ideal minimal over the annihilator L of x; notice that $I \subset L$. Again, E_P is a free module and since $I_P \neq A_P$ we get $\chi(E_P) > 0$, a contradiction.

5.15 *Corollary.* Let E be an A-module with a resolution

$$0 \longrightarrow P_n \longrightarrow \cdots \longrightarrow P_1 \longrightarrow P_0 \longrightarrow E \longrightarrow 0$$

by finitely generated projective modules. Then Ann(Ann(E)) is generated by one idempotent.

Proof: We may define a function $r\colon \mathrm{Spec}(A) \to \mathbb{Z}$, $\mathbb{Z}$ with the discrete topology by $r(P) = \chi(E_P)$. Since $\chi(-)$ is made up of a sum of continuous functions (the rank function of a finitely generated projective module is continuous [11]), the support of r is both open and closed.

On the other hand, the relations above between Ann(E) and the 0th Fitting invariant J of E imply that Ann(Ann(E)) = Ann(J) in each localization. Thus Ann(J) has an open-closed support and is locally 0 or A_P; then Ann(J) is generated by one idempotent [16].

5.16 *Corollary*. Let A be a coherent local ring such that every principal ideal has finite projective dimension. Then A is a domain.

Let A be a coherent domain. (In the following we could avoid the blanket hypothesis of domain by assuming the existence of nonzero divisors in the appropriate places.) In questions of divisibility in A the following set of prime ideals plays a role sufficiently rich on which to base a theory of divisors [55]. $\mathcal{P}$ = all prime ideals minimal over ideals of the type $(a):b = \{r \in A \mid rb \in (a)\}$.

We remark on some of the properties of $\mathcal{P}$.

If I is a fractionary ideal of A, denote the set $\{x \in K \mid xI \subset A\}$, K the field of quotients of A, by I^{-1}. An ideal I is called reflexive if $(I^{-1})^{-1} = I$, that is, if it is reflexive as an A-module. Notice that I^{-1} is always reflexive for any ideal and that $(I^{-1})^{-1}$ is the smallest reflexive ideal containing I.

(a) If I is not contained in any $P \in \mathcal{P}$, $I^{-1} = A$.

(b) If I is a reflexive ideal, then

$$I = \bigcap_{P \in \mathcal{P}} I_P$$

Since (a) is immediate, let us prove (b). Let x be an element in the right-hand side of the expression above. Then $L = I:x = \{r \in A \mid rx \in I\}$ is not contained in any $P \in \mathcal{P}$. Thus $Lx \subset I$ yields $L^{-1} \supset xI^{-1}$. By (a), $L^{-1} = A$; and by a second inversion and the reflexivity of I, we get $x \in I$.

Let $\mathcal{R}$ be the set of finitely generated fractionary ideals I of A having the property that I_P is invertible for each $P \in \mathcal{P}$. On $\mathcal{R}$ we define a composition

$$I \circ J = ((I \cdot J)^{-1})^{-1}$$

That " $\circ$ " is a composition in $\mathcal{R}$ is clear by localization. As for the associativity, just notice that $(I \circ J) \circ L$ and $I \circ (J \circ L)$ are both reflexive, and thus by (b) above it is enough to verify equality at each A_P, $P \in \mathcal{P}$, which is clear.

Note that if one of the ideals above I or J is invertible, then $I \circ J = I \cdot J$, that is, just the ordinary multiplication of ideals.

We now define, essentially in the manner of Ref. 42 (extended to the non-Noetherian case, *e.g.*, in Ref. 52), a function from torsion modules (finitely presented) of finite projective dimension into $\mathcal{R}$.

Let

$$F_1 \xrightarrow{u} F_0 \longrightarrow E \longrightarrow 0$$

be exact, with F_1, F_0 finitely generated projective modules. Let I be the 0th Fitting invariant of E, and define

$$\underline{d}(E) = (I^{-1})^{-1}$$

the divisor of E. That $\underline{d}(E)$ is really in $\mathcal{R}$ results from the next lemma and the comments following (5.13).

5.17 *Lemma*. If $\operatorname{pd}_A E = 1$, then $\underline{d}(E)$ is an invertible ideal of A.

Proof: We prove more generally the following statement. Let E be a finitely generated module over the commutative ring A annihilated by a nonzero divisor. Then $\operatorname{pd}_A E \leq 1$ if and only if I = 0th Fitting invariant of E is a projective (invertible) ideal.

Let

$$0 \longrightarrow K \xrightarrow{u} F \longrightarrow E \longrightarrow 0$$

be exact with F finitely generated and free. If $\operatorname{pd}_A E \leq 1$, K is A-projective. If we localize at any prime, we see that K is free of the same rank as F. Thus K is finitely generated [61]. Hence I is finitely generated and contains a nonzero divisor. To show it is invertible, it is enough to prove it locally. But in this case $I = (\det u)A$.

Conversely, assume that I is invertible. Choosing a partition of the unity [11], we may assume that $I = (d)$. Let $k_1 = (a_{11}, \ldots, a_{1n})$, ..., $k_m = (a_{m1}, \ldots, a_{mn})$ be elements of K such that d is a linear combination of minors of order n extracted from the matrix (a_{ij}). We claim that K is generated by $k_1, \ldots, k_m$ and that it is projective. It will be enough to show that these elements generate K and that K is, at each localization, a free module of same rank as F.

Assume A local; then one of the minors generate the same ideal as (d), say,

$$\det \begin{vmatrix} a_{1,1} & \cdots & a_{1,n} \\ a_{n,1} & \cdots & a_{n,n} \end{vmatrix} = d$$

We show that $k_1, \ldots, k_n$ generate K. Indeed let k be another element of K, and denote by da_i the value of the determinant obtained by substituting the ith row of the matrix above by the row vector k. That

$$k = \Sigma a_i k_i$$

follows then by Cramer's rule.

5.18 *Lemma*. $\underline{d}(-)$ is an additive function on the category of torsion modules of finite projective dimension; that is, if

$$0 \longrightarrow E' \longrightarrow E \longrightarrow E'' \longrightarrow 0$$

is an exact sequence of such modules,

$$\underline{d}(E) = \underline{d}(E') \circ \underline{d}(E'')$$

Proof: If we construct a resolution

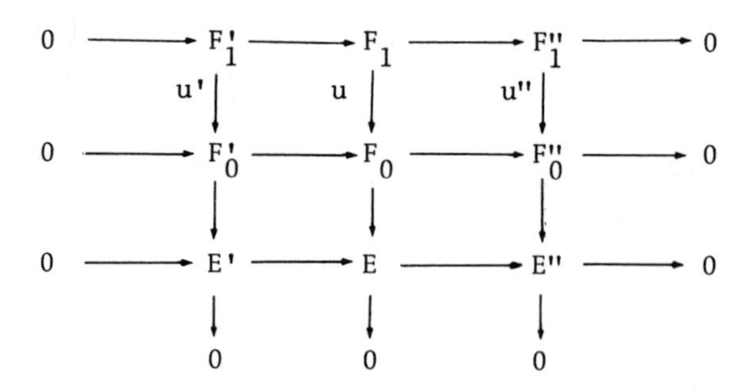

with the F's being free modules, the matrix u is given in terms of u', u'' by

$$u = \begin{vmatrix} u' & * \\ & \\ 0 & u'' \end{vmatrix}$$

In order to show the purported equality, it is enough according to property (b) above to do it at each localization at a prime P in $\mathcal{P}$. But if we do so, we may take u' and u'' to be square matrices and then $\det(u) = \det(u')\det(u'')$.

5.19 *Lemma*. $\underline{d}(E)$ is an invertible ideal of A.

Proof: We may assume that $pd_A E > 1$; let $0 \neq a$ be in the annihilator of E. If

$$F \longrightarrow E \longrightarrow 0$$

is a presentation of E with F free, consider the exact sequence

$$0 \longrightarrow K \longrightarrow F/aF \longrightarrow E \longrightarrow 0$$

Then $pd_A K = pd_A E - 1$. Lemma 5.19 says that

$$\underline{d}(E) \circ \underline{d}(K) = \underline{d}(F/aF) = Aa^n$$

where n is the rank of F. By an induction hypothesis, however, $\underline{d}(K)$ is an invertible ideal and, as remarked earlier, in such cases

$$\underline{d}(E) \circ \underline{d}(K) = \underline{d}(E) \cdot \underline{d}(K)$$

and the conclusion follows.

5.20 *Corollary*. Let I be an ideal of finite projective dimension. Then $I = \underline{d}(A/I)J$, where J is an overdense ideal, that is, $J^{-1} = A$.

Proof: I is the 0th Fitting invariant of the module A/I, and by the definition of $\underline{d}(-)$ we may write $I \subset \underline{d}(A/I)$. We can express I as a product $\underline{d}(A/I)J$. Taking inverses twice, we get

$$(I^{-1})^{-1} = \underline{d}(A/I)(J^{-1})^{-1} \qquad \text{or} \qquad (J^{-1})^{-1} = A$$

But then $((J^{-1})^{-1})^{-1} = A$ and $J^{-1} = A$ as desired.

We may now state the main implication of these considerations.

5.21 *Theorem*. Let A be a local coherent ring in which every finitely generated ideal has a finite projective resolution. Then A is a GCD domain.

Proof: Corollary 5.16 says that A is a domain. Let a, b be two nonzero elements of the maximal ideal M of A, and write $I = (a,b)$. Applying the above decomposition to I, we get

$$(a,b) = \delta(\alpha,\beta) \qquad \text{where} \quad (\alpha,\beta)^{-1} = A$$

Consider the sequence

$$0 \longrightarrow K \longrightarrow A^2 \xrightarrow{\phi} (\alpha,\beta) \longrightarrow 0$$

with $\phi(x,y) = x\alpha - y\beta$. If $(x,y) \in K$, that is, $x\alpha = y\beta$, $x/\beta \in (\alpha,\beta)^{-1} = A$ and hence $x \in A\beta$. Thus K is generated by (β,α), and the conclusion follows.

This is essentially MacRae's theorem [41] that says that in a Noetherian ring every two-generated ideal has projective dimension 0, 1, or ∞. In particular, it provides a generalization of the unique factorization of regular local rings [4].

5. *Local rings of Tor-dimension two*

Here we consider various aspects of the theory of coherent local rings of Tor-dimension two and contrast them to similar facets of rings of global dimension two.

First consider the case of a coherent local ring A with maximal ideal M finitely generated and of finite projective dimension. According to (5.11) every finitely generated ideal has bounded projective dimension and Tor-dim(A) = pd(A/M).

These rings display a feature found in the Noetherian regular local rings: that its maximal ideal is generated by a regular sequence [2]. Indeed, if $x_1, \ldots, x_n$ is a minimal generating set for M, as in Ref. 35, we get a decomposition

$$M/x_1M = (x_1)/x_1M \oplus M/(x_1)$$

If

$$0 \longrightarrow F_r \longrightarrow \cdots \longrightarrow F_1 \longrightarrow F_0 \longrightarrow M \longrightarrow 0$$

is a free resolution of M, tensoring with $A/(x_1)$ we conclude (since x_1 is a nonzero divisor) that

$$0 \longrightarrow F_r/x_1F_r \longrightarrow \cdots \longrightarrow F_1/x_1F_1 \longrightarrow F_0/x_1F_0 \longrightarrow M/x_1M \longrightarrow 0$$

is exact.

Thus M/x_1M has finite projective dimension over the ring $A/(x_1)$ and, from the decomposition above its maximal ideal, $M/(x_1)$, also has finite projective dimension. Thus we are back in the preceding situation but with a "smaller" maximal ideal.

5.22 *Theorem.* Let {A,M} be a local coherent ring with M finitely generated and $d = pd_A M < \infty$. Then Tor-dim(A) = d + 1, and M is generated by a regular sequence of d + 1 elements.

5.23 *Remark.* It is, however, no longer true that all maximal regular sequences in M have the same length d + 1.

Let A be the ring

$$Q[x,y]_{(x,y)}[[t))$$

considered in Ref. 63. A has Tor-dimension two and its maximal ideal is (x,y). Notice, however, that {t} is a maximal regular sequence in M also.

For the remainder of this section, let A be a coherent local ring of Tor-dimension two. If M is its maximal ideal, one of the facts that we considered in the description of the local rings of global dimension two was the dimension of the vector space M/M^2 over A/M. Call this number k.

We found three possibilities for k: 0, 1 and 2, the first two corresponding to valuation domains. Although this identification no longer holds, we still have the following proposition.

5.24 *Proposition.* $k \leq 2$.

Proof: Suppose x_1, x_2, x_3 are three elements of M, with one of them not in M^2. If they are linearly independent modulo M, a minimal resolution of the ideal I they generate has the form

$$0 \longrightarrow A^2 \xrightarrow{u} A^3 \longrightarrow I \longrightarrow 0$$

By Burch's theorem I would then be generated by the minors of order two of the matrix u and would lie in M^2.

5.25 *Corollary.* If $k = 2$, then M is generated by two elements.

Proof: The argument above immediately adapts, for $k = 2$, to show that M is finitely generated.

In the cases $k = 0$ or 1, A is not necessarily a valuation domain as the following example shows.

5.26 *Example.* Let {V,N} be a valuation domain with its maximal ideal N not finitely generated. Let $A = V[x]_{(x,N)}$. By the syzygies' theorem 0.14, Tor-dim(V[x]) = 2, whereas by Ref. 26, V[x] is coherent. Thus A is coherent and Tor-dim(A) ≤ 2. But its maximal ideal $M = (x,N)A$ is such that $k = 1$, but it is not a valuation ring.

In the case $k = 1$, this feature is always present.

5.27 *Proposition.* If $k = 1$, there is an element x in M such that $V = A/(x)$ is a valuation domain.

Proof: Let $x \in M \setminus M^2$. Since A is a GCD domain [see (5.21)], x is a prime element and V is a domain. Let b, c be elements of M; we must show that, modulo (x), one of them is a multiple of the other. For that it is enough to prove that the ideal (x,b,c) is minimally generated by two elements, which we know is the case from (5.24).

Finally, the case $k = 0$ does not correspond to a valuation domain.

5.28 *Example.* Let F be a field, and consider the sequence of polynomial rings

$$F[x,y] \subset F[x^{1/2},y^{1/2}] \subset F[x^{1/4},y^{1/4}] \subset \cdots$$

and let B be its union. Note that each ring is free as a module over the preceding; B is then a coherent ring of Tor-dimension 2. Let A be the localization of B at the "origin." A has Tor-dimension two and a maximal ideal $M = M^2$.

6. *Difficulties in higher dimensions*

A first difficulty in attempting a description of the local rings of global dimension three or higher is the possible lack of coherence. In dimension three, for instance, it is no longer true that domains are coherent. A feature of the two-dimensional case is still present however.

5.29 *Theorem.* Let $\{A,M\}$ be a coherent local ring of finite global dimension. If $\text{gl dim}(A) = \text{Tor-dim}(A)$, then M is finitely generated.

Before we embark on its proof, notice that from (5.22) M will be generated by gl dim(A) elements.

If A is a coherent local ring of global dimension three, the possibilities predicted by (5.29) are:

(1) Tor-dim(A) = 1, and A is a valuation domain.

(2) Tor-dim(A) = 3, and M is generated by a regular A-sequence of three elements.

(3) Tor-dim(A) = 2; in this case M can be either non-finitely generated or minimally generated by two elements.

Proof: Let E be a finitely generated module of maximum projective dimension [= n = Tor-dim(A) = gl dim(A)]. Let

$$0 \longrightarrow F_n \xrightarrow{\phi_n} F_{n-1} \longrightarrow \cdots \longrightarrow F_1 \xrightarrow{\phi_1} F_0 \longrightarrow E \longrightarrow 0$$

be a projective resolution of E with F_i finitely generated. If $\text{rank}(F_n) = r$, let I be the ideal of A generated by the minors of order r of the matrix that represents ϕ_n.

Claim: $\text{Ext}_A^i(A/I,A) = 0$, $i < n$.

Let $B = A[x_1,x_2,\ldots]$ be the polynomial ring in infinitely many indeterminates over A. B has the following property: If J is a finitely generated ideal of B, then in the ring $C = B/J$ any finitely generated faithful ideal contains a nonzero divisor.

To sustain the claim, consider

$$\text{Ext}_A^i(A/I,A) \;{}_A\, B \simeq \text{Ext}_B^i(B/IB,B)$$

a valid isomorphism since A/I has a finite infinite presentation and B is A-flat. Change the notation, and assume we have a resolution as above over a ring A with the stated property of B. (Warning: We are not assuming that B is coherent.)

Since ϕ_n is injective, by McCoy's theorem there is an element f_1 in I that is a nonzero divisor. Tensor the sequence

$$0 \longrightarrow F_n \xrightarrow{\phi_n} F_{n-1} \longrightarrow \cdots \longrightarrow F_1 \longrightarrow \operatorname{im}(\phi_1) \longrightarrow 0$$

by $A^* = A/(f_1)$ to get the exact sequence

$$0 \longrightarrow F_n^* \longrightarrow F_{n-1}^* \longrightarrow \cdots \longrightarrow F_1^* \longrightarrow \operatorname{im}(\phi_1)^* \longrightarrow 0$$

5.30 *Lemma*. Let A be a ring, E an A-module, and f an element that is a nonzero divisor with respect to both A and E. We then have the isomorphism

$$\operatorname{Ext}_A^{j+1}(-,E) \simeq \operatorname{Ext}_{A/(f)}^{j}(-,E/fE)$$

in the category of $A/(f)$-modules.

Proof: See Ref. 36.

In the situation above we get

$$\operatorname{Ext}_A^{i}(A/I,A) \simeq \operatorname{Ext}_{A/(f_1)}^{i-1}(A/I,A/(f_1))$$

An appropriate induction hypothesis allows the conclusion in the claim.

To complete the proof, notice that the pair of rings $\{A,A/I\}$ is admissible for the change of rings and dimensions described in (5.1). A/I is then a ring of finitistic projective dimension zero. Since it is also coherent, A/I is Artinian and M is indeed finitely generated.

Chapter 6

FINITELY GENERATED IDEALS

In this chapter we begin a study of the rings of global dimension two that are not necessarily local. The approach we employ is double-edged: (i) use of the local classification theorem [(2.2) and (4.10)] and (ii) use of the change of rings results of Chapter 5 to examine the nature of the rings A/I, I an ideal of A. The restriction, however, of finite generation of I (see, *e.g.*, Theorem 5.7) places natural limitations on this method, and the analysis will not always be complete. Nevertheless, for some aspects, in particular those concerned with the question on whether a given prime ideal is finitely generated--and this will appear often in a ring which is not Prufer--the method seems sufficiently rewarding to warrant the effort.

1. Coherence

If the strategy outlined above is going to meet with any success, we must have a reasonably simple description at least of the finitely generated ideals.

Let A be a ring of global dimension two and I a finitely generated ideal. If we write a resolution

$$0 \longrightarrow K \xrightarrow{u} A^n \longrightarrow I \longrightarrow 0$$

K will be a projective module. With K finitely generated we may, by an appropriate partitioning of the spectrum of A, assume K A-free and relate I to the matrix u between the two free modules.

Another need for the finite generation of K arises if we want to use the local classification theorem. As it is often the case, I will be involved in some Hom functor and the localization procedure will work more smoothly (if at all) under the hypothesis of finite presentation of I.

Thus the first question to consider is that of the coherence of A. This seems to be the first natural divider of the general theory since it is closely related to the topology of Spec(A) with the coherent case much more amenable.

In this section let A be a ring of global dimension two. If P is a prime ideal, the localization A_P has global dimension at most two. Some elementary consequences of this are:

(1) A is a PF-ring, *i.e.*, every principal ideal is flat.

(2) A is reduced.

(3) Every prime ideal contains a unique minimal prime.

(4) If P is a minimal prime ideal, P is pure and A/P has global dimension at most two.

If A is coherent, for each element $f \in A$ the flat ideal Af is projective since it has a finitely generated annihilator [60]. Conversely, suppose that A is a PP-ring, and let I be a finitely generated ideal, say, $I = (a_1,\ldots,a_n)$. If $Ae_i = 0:Aa_i$ (e_i = idempotent), then $0:I = Ae$, $e = e_1 \cdots e_n$. By taking a decomposition $A = Ae \times A(1 - e)$, we may assume that I is an ideal of the ring of global dimension two $A(1 - e)$ [since by (2), $Ae \cap I = 0$].

If we change notation and write

$$0 \longrightarrow K \longrightarrow A^n \longrightarrow I \longrightarrow 0$$

with $0:I = 0$, now K is a projective module which is a free module of rank $n - 1$ at each localization; K is then finitely generated [61].

Taking into account the characterization of PP-rings (Proposition 1.13) in terms of the minimal spectrum, we may write, more suggestively, the following proposition.

6.1 *Proposition*. A is coherent if and only if Min(A) is compact.

We have already seen an example (Example 1.3b) of a noncoherent ring of global dimension two (the annihilator of 2A is not finitely generated). Another example, this time without nontrivial idempotents, will now be given.

6.2 *Example* (Jensen). Let M be a family of pairwise disjoint intervals of the real line with rational endpoints, such that between any two intervals of M there is at least another interval in M. Let A be the ring of continuous functions that are rational constant except on finitely many of these intervals on which it is given by a polynomial with rational coefficients. Then every finitely generated ideal of A is principal (left as an exercise!), A is reduced, and hence Tor-dim(A) = 1. Since A is a countable ring, by Lemma 0.11, gl dim(A) $\leq$ 2. A is noncoherent: If f is given by

$$f(x) \begin{cases} 1 - x & 0 \leq x \leq 1 \\ 1 + x & -1 \leq x \leq 0 \\ 0 & \text{elsewhere} \end{cases}$$

then 0:f is not finitely generated. Also, A is indecomposable.

6.3 *Remark*. (1) We saw in Theorem 1.13 that if A is a PP-ring, then so is A[x]. Thus, if A is a hereditary ring, A[x] is a coherent ring of global dimension two.

(2) Domains of global dimension two are coherent. Generalizing the local case, let A be such that its Jacobson radical J is a prime ideal. Then A is a domain. To see this, let $0 \neq f \in A$; tensor

$$0 \longrightarrow 0{:}f \longrightarrow A \longrightarrow Af \longrightarrow 0$$

by A/J to get

$$0 \longrightarrow (0{:}f)/J(0{:}f) \longrightarrow A/J \longrightarrow Af/Jf \longrightarrow 0$$

But (0:f)/J(0:f) is a pure ideal of the domain A/J and thus = 0 (since $Af \neq Jf$ by Nakayama's lemma). By localization at the maximal ideals, it then follows 0:f = 0.

6.4 *Remark*. If A is a ring of global dimension two and I is a finitely generated faithful ideal, then A/I is coherent.

Indeed, let J be a finitely generated ideal containing I. J being faithful, is of finite presentation. The sequence

$$0 \longrightarrow I \longrightarrow J \longrightarrow J/I \longrightarrow 0$$

then says (by Schanuel's lemma) that J/I has finite presentation as an A-module. Thus there is an exact sequence

$$A^m \longrightarrow A^n \longrightarrow J/I \longrightarrow 0$$

which by tensorization with A/I gives the desired conclusion.

2. *Burch's theorem revisited*

Let I be a finitely generated faithful ideal over the commutative ring A and assume pd I = 1. Write a resolution

$$0 \longrightarrow K \xrightarrow{u} A^n \longrightarrow I \longrightarrow 0$$

(K is finitely generated!) and write D for the ideal generated by the "minors" of order $n - 1$ of u (that is, let D be the Fitting invariant of order one of I). Let A be a PF-ring [*i.e.*, if gl dim(A) = 2].

6.5 *Theorem.* $I = \text{Hom}(D,I) \cdot D$, where Hom(D,I) is a projective ideal of A and D is overdense.

Observe that this is an explicit representation of I which should be compared to Corollary 5.20. We have already stated it in its original form as Theorem 2.3.

Proof: Pick a partition of the unity in the manner of Ref. 11 to make K free on each open set. Let $a_1, \ldots, a_n$ be the generating set for I, and write $u = (c_{ij})$. The relations

$$\Sigma c_{ij} a_j = 0 \qquad i = 1, \ldots, n - 1$$

give (Cramer's rule!) that

$$d_k a_j = d_j a_k$$

where d_k is the minor obtained by leaving out the kth column of the matrix u. Define the following map:

$$\psi\colon D \longrightarrow I \qquad \text{by} \quad \psi(\Sigma r_j d_j) = \Sigma r_j a_j$$

First we note that ψ is well-defined: $\Sigma r_j d_j = 0$ yields

$$a_k(\Sigma r_j d_j) = (\Sigma r_j a_j) d_k = 0 \qquad \text{and} \qquad (\Sigma r_j a_j)D = 0$$

By McCoy's theorem, D is a faithful ideal, and hence $\Sigma r_j a_j = 0$ also.

Thus ψ is an isomorphism, and D also has projective dimension one and is of finite presentation.

Now we show that the natural homomorphism

$$\phi\colon A \longrightarrow \text{Hom}(D,A) \qquad \phi(a)(d) = ad$$

is an isomorphism.

From the remark above about D, we may localize at a prime ideal and thus assume A to be a domain.

Hom(D,A) may be identified with $\{x \in \text{quotient field of } A \mid xD \subset A\}$. Let $a/b \in \text{Hom}(D,A)$; if we tensor the presentation above of I by A/(b), we get (recall $K = A^{n-1}$)

$$0 \longrightarrow (A/bA)^{n-1} \longrightarrow (A/bA)^n \longrightarrow I/bI \longrightarrow 0$$

still exact. Using McCoy's theorem again, the Fitting invariant of order one of I/bI is still faithful, and thus $a \in Ab$.

This implies that the map above is actually given by multiplication by an element of A. That Hom(D,I) is projective follows now since it is locally (in the Zariski topology) generated by a nonzero divisor.

6.6 *Remark*. Gruson has suggested a way that shows the decomposition of (6.5) to be valid in all cases, not just when, as we have been explicitly assuming, I contains a nonzero divisor. The key fact in the decomposition is to show Hom(D,A) = A, which can be done by changing the ring to A[t]: The ideal I of A being faithful, IA[t] contains a nonzero divisor (say the polynomial $a_n t^n + \cdots + a_1 t$) and the above argument works. The change of rings being faithfully flat, the descent is permissible.

3. The Δ-theorem

In this section we consider conditions that ensure that a prime ideal in a ring A of global dimension two is finitely generated. This will be the case if the ring is coherent and the prime ideal contains two noncomparable primes.

As a corollary it will follow that a Krull domain of global dimension two is Noetherian. Another amusing consequence is that if A is coherent and not semihereditary, it contains a finitely generated maximal ideal [57].

The main result here is the following theorem.

6.7 *Theorem*. Let A be a coherent ring of global dimension two, and let M be an ideal of A containing two noncomparable prime ideals P and Q. Then M is finitely generated.

The key technical element in the proof is contained in the change of rings theorem of (5.7).

6.8 *Lemma*. Let A be a ring of global dimension two. If I is an overdense finitely generated ideal of A, then A/I is an Artinian ring. In particular, the ideals above I are finitely generated.

Proof: Consider the ring A/I: Because of (5.7), the finitistic projective dimension of A/I is zero. Since I is faithful, by (6.4) the ring A/I is coherent. It then follows from Proposition 1.11 that A/I is an Artinian ring.

Proof of (6.7): Since P and Q are noncomparable prime ideals, let us pick elements $a \in P \setminus Q$ and $b \in Q \setminus P$. Let $0:a = Ae$, $0:b = Ag$, with e, g idempotents; this is possible since A is coherent. Notice that $e \in Q$ and $g \in P$; thus the ideal I generated by a, b, e and g is contained in M. Resolve this ideal in the following way:

$$0 \longrightarrow L \longrightarrow A^4 \xrightarrow{h} I \longrightarrow 0 \qquad (*)$$

where $h(x,y,z,w) = ax + by + ez + gw$. Since gl dim(A) = 2, $\mathrm{pd}_A I \leq 1$; and thus L is projective. L is also finitely generated, for A is coherent. Since L is a finitely generated projective module, according to Ref. 11 there is a partition of the unity $(f_1,\ldots,f_n) = A$ with the property that L_f is A_f-free of finite rank. (here f stands for any of the f_i's and L_f is the module of fractions arising out of the multiplicative set of powers of f.) Again by Ref. 11, we can prove M is finitely generated by proving M_f is finitely generated in A_f for those $f \notin M$. Notice that in these cases the choices of a, b, e, and g are not affected. Thus change notation and assume that L is a free A-module of rank 3. Thus, (*) becomes the following exact sequence

$$0 \longrightarrow A^3 \longrightarrow A^4 \longrightarrow I \longrightarrow 0$$

Note that $a + e$ is a regular element in I. By Theorem 6.5 we have $I = tD$, where $D = (d_1,d_2,d_3,d_4)$, t regular, and $D^{-1} = A$. It is easy to see $d_1 + d_3$ is a regular element in D; hence, by (6.5), D is overdense. As (6.8) indicated, if we show $D \subset M$, then M will be finitely generated. To check this, let us first note that $e \in Q$, $g \in P$, and $a = td_1$, $b = td_2$, $e = td_3$, $g = td_4$. Clearly $t \notin P$ or $t \notin Q$, since otherwise $a = td_1 \in Q$ or $b = td_2 \in P$, which yields a contradiction. Since P is a prime ideal and $a = td_1 \in P$, we have $d_1 \in P \subset M$. Similarly, $d_4 \in P$ and d_2,d_3 lie in $Q \subset M$.

That the lower "vertices" of the triangle are not necessarily finitely generated follows from the next example.

6.9 *Example*. Let Q be the field of rationals, and consider the two subrings of $Q(x,y)$:

$$V = Q[x,y]_{(x,y)}[xy^{-n},\ n = 1, 2, \ldots]$$

and

$$N = Q[x,y]_{(x,y-1)}$$

V is a valuation domain of Krull dimension two with a principal maximal ideal, and N is a two-dimensional regular local ring.

Let $A = V \cap N$; the spectrum of A looks like

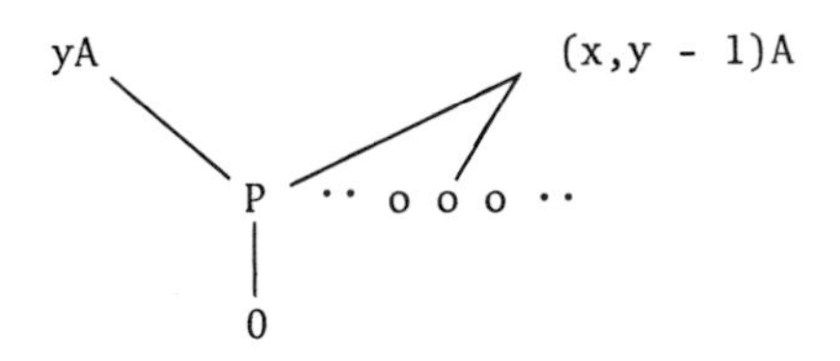

Thus A has global dimension two and $P = A(xy^{-n}, n = 1, 2, \ldots)$ is the only non-finitely generated prime ideal.

If A is a ring of global dimension two, its Tor-dimension (as we have seen by examples) might be 0, 1, or 2. If Tor-dim(A), and A is coherent, A is semihereditary [18]. The following "existence" result applies if the global dimension is "strictly" two.

6.10 *Corollary*. If A is a coherent ring and gl dim(A) = Tor-dim(A) = 2, there is a finitely generated maximal ideal.

Proof: The condition that Tor-dim(A) = 2 amounts to saying, according to Proposition 1.2, that for some prime ideal P, A_P is not a valuation domain. Since A_P is of global dimension two, it must, by Theorem 2.2, be either a two-dimensional Noetherian ring or an umbrella ring. In either case we have P maximal containing two noncomparable primes and (6.6) applies.

We recall from Ref. 55 that a domain A is a Krull ring if it satisfies the following two conditions: (1) If P is a prime ideal of height one, then A_P is a Noetherian valuation domain. (2) An arbitrary principal ideal aA of A is the intersection of a finite number of primary ideals whose associated primes are of height one.

6.11 *Corollary*. A Krull domain A of global dimension two is Noetherian.

Proof: Let M be a prime ideal of height > 1. A_M has then global dimension two, and it is also a Krull domain [55]. Since the prime ideals minimal over a principal ideal are of height one, Theorem 2.2 implies that A_M is Noetherian of Krull dimension two. Using (6.7) now, we get M finitely generated.

What is left is to examine the prime ideals of height one. Let P be one such. A_P is a discrete valuation domain with a maximal ideal generated by, say, $a \in P$. Let

$$aA = Q \cap Q_1 \cap \cdots \cap Q_n$$

be the primary decomposition of aA, with Q its P-component. Taking into account the choice of a, $Q = P$. Choose now $b \in Q_1 \cap \cdots \cap Q_n \setminus P$. Then

$$aA:b = (P:b) \cap (Q_1:b) \cap \cdots \cap (Q_n:b) = P$$

Since A is coherent, aA:b is a finitely generated ideal.

4. *Generating prime ideals efficiently*

Let P be a prime ideal of a ring A of global dimension two. If P is finitely generated, we show that P will be generated by the same number of generators one obtains with A Noetherian [59].

We begin by examining the nature of the ring A/P, for P finitely generated. First we note that if P is not faithful, it must be a minimal prime ideal; otherwise its annihilator I would be contained in each minimal prime and would be a nilideal. Since A is reduced, $I = 0$.

Thus if $I \neq 0$, P will be generated by one idempotent since it is a pure ideal. Let us assume $I = 0$.

P admits a resolution

$$0 \longrightarrow K \longrightarrow A^n \longrightarrow P \longrightarrow 0$$

with K finitely generated projective. If P is not projective, for some maximal ideal M containing P we have

$$\mathrm{Ext}^1_A(P,K)_M = \mathrm{Ext}^1_{A_M}(P_M,K_M) \neq 0$$

Thus P_M is a finitely generated prime ideal of A_M of projective dimension one. Theorem 2.2 then says that P must be equal to M.

6.11 *Theorem*. Every finitely generated prime ideal P of a coherent ring A of global dimension two can be generated by three elements.

Proof: As remarked earlier, we may assume that P is not a minimal prime.

Case (i): pd $P = 0$.

P/P^2 is then a projective module of rank one over the Noetherian ring A/P of Krull dimension one. It can then be generated by two elements (see Theorem 1.16). It follows that P, by (1.14), can be generated by three elements.

Case (ii): pd $P = 1$.

Now we use the hypothesis that A is coherent. As seen above, A_P is not a valuation ring, and thus its maximal ideal can be generated by two elements $a,b \in P$. Let $Ae = 0:a$, e an idempotent. Notice that $e \in P$ and that $(a + e,b)_P = (a,b)_P = PA_P$. We may then assume that (a,b), by changing notation, generate P at A_P and that a is a nonzero divisor.

Decompose (a,b) according to (6.5) as

$$(a,b) = JL$$

with J projective and L overdense. $J \not\subset P$ since otherwise PA_P would be principal. Let $P, M_1, \ldots, M_n$ be the maximal ideals containing L. We have

$$P + J \cdot M_1 \cdots M_n = A$$

and thus a relation

$$c + d = 1 \qquad \text{with } c \in P \text{ and } d \in J \cdot M_1 \cdots M_n$$

We claim $(a,b,c) = P$. It suffices to check at the localizations. Let Q be a maximal ideal. If $(a,b) \not\subset Q$, we are done. If $J \subset Q$, $c \not\in Q$ and again we conclude that (a,b,c) blows up at Q. Finally, if $L \subset Q$, Q must be one of the ideals $P, M_1, \ldots, M_n$. If $Q = P$, the choices of a, b take over, while in the other cases c lies outside of Q. This completes the proof.

It is likely that the coherence of A is redundant for the validity of case (ii).

5. *Noetherianization*

In section 3 we saw that a Krull domain of global dimension two must be Noetherian. Also, the example of Chapter 2 gives the standard use of the local classification theorem to verify (by comparison) whether a local ring has global dimension two.

Here we discuss two conditions that make a ring of global dimension two Noetherian. They are perhaps the conditions more immediately related to the Noetherianess of a ring A.

(SN) Spec(A) is a Noetherian space.

(LN) Every localization of A at a prime ideal is Noetherian.

6.13 *Theorem.* If a ring A of global dimension two satisfies (SN) and (LN), it is Noetherian.

Actually, it is likely that (SN) is too strong. In fact, we shall proceed under the hypothesis that the maximal spectrum be Noetherian until the difficulty becomes (apparently) insurmountable.

First we consider the following lemma.

6.14 *Lemma*. Let A be a commutative ring with Max(A) Noetherian. Then each pure ideal of A is generated by one idempotent.

Proof: Let I be a pure ideal of A. Since Max(A) is Noetherian, there is a finitely generated ideal $J \subset I$ such that $V(I) = V(J)$ in Max(A). Localizing at maximal ideals easily checks that J is also a pure ideal; hence $J = Ae$, $e = e^2$. It is clear that $I = Ae$ also.

6.15 *Proposition*. Let A be a ring of global dimension two with Max(A) Noetherian. Then A is a finite product of domains; in particular A is a coherent ring.

Proof: It is enough to show that there are only finitely many idempotents in A. Indeed, if A is connected and P is a minimal prime, since P is pure by the preceding lemma it is generated by one idempotent, that is, $P = 0$.

If e is an idempotent of A, consider the decomposition

$$A = Ae \times A(1 - e)$$

Arguing with the decomposability of the factors, we obtain that A either has only finitely many idempotents or admits a sequence

$$(e_1) \subset (e_2) \subset \cdots$$

of idempotents $\neq 1$. The ideal I they generate is then a non-finitely generated pure ideal.

Proof of (6.13): We may assume that A is a domain. Because A is locally Noetherian, its Krull dimension has to be at most 2. We use Cohen's theorem; let P be a prime ideal.

Case (i): height(P) = 2.

P is then a maximal ideal and the Δ-theorem then says that it is finitely generated. More elementarily, because Max(A) is Noetherian, P is the radical of a finitely generated idealI; add to I a finitely generated subideal J of P which generates P at A_P. Then $P = I + J$.

Case (ii): height(P) = 1.

If P is maximal, the proof is much the same as above. Suppose thus that A/P is a domain of Krull dimension one. Notice it is a Noetherian ring by the previous case. There two subcases to consider depending on the finiteness of Spec(A/P).

Case (iia): Spec(A/P) is finite.

Again, since Max(A) is Noetherian, $V(P) = V(I)$, where I is a finitely generated subideal of P. Let $V(P) = \{M_1, \ldots, M_n\}$; choose $a_i \in P$ such that $(a_i)_{M_i} = P_{M_i}$ for each i (actually, we may find a single a that works in all cases). It follows easily that

$$P = (I, a_1, \ldots, a_n)$$

Case (iib): Spec(A/P) is infinite.

As before, we may write $V(P) = V(I)$, I a finitely generated subideal of P. Choose $a \in P$ such that $aA_P = PA_P$. Write $I' = (I, a)$. The spectrum of A/I' looks like

... o o o ...

... P o o ...

where the top row consists of the maximal ideals containing P, whereas the bottom row are the minimal primes of I'. Observe that each prime ideal Q minimal over I' and distinct from P is such that it is contained in only finitely many maximal ideals, and is, by the preceding case, finitely generated. This pathology is what makes it likely that Max(A) = Noetherian suffices to prove (iib).

We now complete the proof under the (stronger) original hypothesis, that Spec(A) = Noetherian.

In this case the number of prime ideals minimal over I' is finite, say, P, $Q_1, \ldots, Q_n$. Choose $b \in P \setminus \cup Q_i$, and write

$$I'' = (I', b)$$

Using the decomposition provided by (6.5),

$$I'' = J \cdot L$$

with J projective and L overdense. Thus $L \not\subset P$, and we claim finally $P = J$. Notice that P is the only prime ideal minimal over J. To show the desired equality, it is enough (by the choice of I) to check the localizations at maximal ideals containing P. A_M being a regular local ring, it is a UFD. We may then write $J_M = P^r A_M$ since P_M is the only prime ideal minimal over the principal ideal J_M. By the choice of a above we conclude that $r = 1$, which completes the proof.

Chapter 7

DOMAINS

In order of complexity the rings of global dimension two could be diagrammatically divided into:

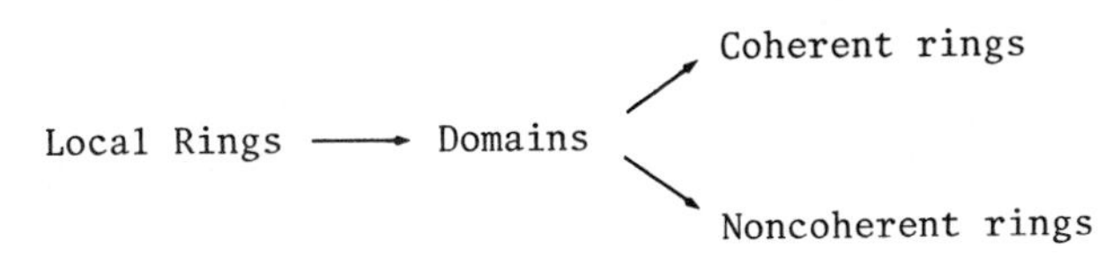

The study of domains, still in a highly embryonic state, consists so far in following the basic theme running through these notes: To what extent is a domain of global dimension two made up of the basic constituents, Prüfer domains and Noetherian rings. In section 1 of this chapter we discuss an infinity of conditions of a local and residual nature on a domain of global dimension two that might make it more recognizable. By contrast section 2 uses *ad hoc* methods to give a fairly complete description of the symmetric algebras of global dimension two. In section 3 we show that coherent rings of global dimension two admit a sheaf representation as a ringed space over a compact topological space with domains of global dimension two as stalks. As for noncoherent rings, the situation is pretty dark.

1. $\aleph_0$-Noetherian domains

In this section we examine the domains of global dimension two in terms of an infinity of local conditions to take advantage of Theorem 2.2 and lower dimensional residual rings.

Specifically, we consider the consequences of the following properties of a domain A.

(X) Every localization A_P is a ring of global dimension ≤ 2.

(Y) For every $0 \neq a \in A$, $FPD(A/aA) \leq 1$.

At the very end we will need a cardinality hypothesis of the following type:

(Z) A is χ_0-Noetherian.

The local rings of global dimension two are χ_0-Noetherian as shown in Theorem 4.8. In Example 4.28 we saw, however, a ring of global dimension two that was not χ_0-Noetherian.

The following proposition is a first consequence of conditions (X) and (Y).

7.1 *Proposition.* If a domain A satisfies (X) and (Y), it is coherent.

Proof: Let I be a nonzero finitely generated ideal of A, and write

$$(*) \qquad 0 \longrightarrow L \longrightarrow A^n \longrightarrow I \longrightarrow 0$$

a projective presentation of I. Because of (X), since local rings of global dimeension two are coherent, for each prime P, L_P is a free A_P-module of rank $n - 1$. Let a be a nonzero element of I; tensoring the sequence above by A/aA, we get

$$0 \longrightarrow L/aL \longrightarrow (A/aA)^n \longrightarrow I/aI \longrightarrow 0$$

Since L/aL is a flat A/aA-module, it follows from Lemma 0.11 that (Y) that L/aL is a projective A/aA-module of constant rank $n - 1$. According to Ref. 61, L/aL is then a finitely generated A/aA-module. Say L_1 is a finitely generated submodule of L such that $L = L_1 + aL$.

On the other hand, let S be the multiplicative set of powers of a, and tensor the sequence (*) by A_S to conclude that L_S is a free A_S-module. Let L_2 be a finitely generated submodule of L such that $L_S = (L_2)_S$.

We claim that $L = L_1 + L_2$. But this is immediate by localization and Nakayama's lemma.

7.2 *Proposition.* Let A be a domain with (X) and (Y). If I is a finitely generated overdense ideal, then A/I is an Artinian ring.

Proof: Let $I = (a_1, \ldots, a_n)$, and consider the ideals

$$I_{ij} = (a_i, a_j)$$

According to (6.5), we may write $I_{ij} = P_{ij}J_{ij}$, with P_{ij} projective and J_{ij} overdense. Let us now consider a partition of the unity such that all P_{ij} become principal, that is, in each corresponding open set $\mathrm{Spec}(A_f)$,

$$(a_i, a_j) = (d_{ij})J_{ij}$$

where d_{ij} is the GCD of a_i, a_j.

Let M be a prime ideal containing I. We have only to show that, in $\mathrm{Spec}(A_f)$, $V(I_f)$ is a finite set. If some $d_{ij} \notin M$, J_{ij}, which is an ideal generated by a regular sequence of two elements, is contained in M. From (Y), however, $\mathrm{FPD}(A/J_{ij}) = 0$ for that particular ideal. Since the ring A is coherent, it then follows that A/J_{ij} is Artinian.

So assume that no J_{ij} is contained in M. Let us then localize at M. A_M is a GCD domain, and the hypothesis $d_{ij} \in M$ contradicts that the ideal $(a_1,\ldots,a_n)_M$ is overdense.

We are now ready to prove the main consequence of (X) and (Y). It will mimic Gruson's theorem 0.16 and uses the techniques developed in Ref. 26.

Let I be an ideal of a ring A with (X) and (Y), and let

$$0 \longrightarrow L \longrightarrow F \longrightarrow I \longrightarrow 0$$

be a projective presentation of I.

7.3 *Proposition.* L is a Mittag-Leffler module.

Proof: By (X) we know that for each prime P, L_P is a free A_P-module. We use (2.5.6) in Ref. 26(Part II) to see that L is a Mittag-Leffler module. Then let

$$u: G \longrightarrow L$$

be a map from the finitely presented A-module G into L which is universally injective at a prime P. In particular G_P is a free A_P-module. By a previous localization we may assume that G is a free A-module and that u is injective. Following the argument of (3.2.6) of Ref. 26(Part II), let v be the composition of u with the inclusion of L in F (which may be assumed free). Let J be the determinant of v, that is, if G is a free module of rank r, det v is the ideal generated by the coordinates of $\mathrm{im}(\Lambda^r v)$ in $\Lambda^r F$. If $J \not\subset P$, choose $f \in J \setminus P$, and v_f (and hence u_f) is universally injective. In any case $D(f) \subset U$, the set of primes where u is universally injective. Tensor the presentation of I with A/fA to get

$$0 \longrightarrow \mathrm{Tor}_A^1(C,A/fA) \longrightarrow G/fG \xrightarrow{u \quad A/fA} L/fL \longrightarrow C/fC \longrightarrow 0$$

where C denotes the cokernel of u. Since

$$L/fL \hookrightarrow F/fF$$

$$\ker(u \quad A/fA) = \ker(v \quad A/fA)$$

On the other hand, since A/fA is coherent by (7.1), $K = \mathrm{Tor}^1_A(C, A/fA)$ is a finitely generated submodule of G/fG. If we localize at P, we get $K_P = 0$ since u_P is universally injective. We may then, by passing to some D(g), $g \notin P$, assume that K = 0, that is, we have the inclusion

$$0 \longrightarrow G/fG \xrightarrow{\bar{u}} L/fL$$

As remarked earlier, L/fL is A/fA-projective, and thus there is an open set on which $\bar{u}$ is universally injective. We do this reduction to a still smaller open set of Spec(A) and obtain

(1) C/fC is A/fA-flat.

(2) $\mathrm{Tor}^A_1(C, A/fA) = 0$

(3) f is a nonzero divisor.

We then fall into the conditions of (1.4.2.1) of Ref. 26(Part II) and conclude that C is A-flat, and thus u is universally injective.

If we now add the hypothesis (Z), we could take F countably generated, and thus L would be a countably generated, flat, Mittag-Leffler module, that is, a projective module.

7.4 *Theorem.* If A is a domain satisfying conditions (X), (Y), and (Z), it has global dimension two.

It is likely that (X) and (Y) already suffice to ensure gl dim(A) = 2. Further evidence of this is provided by the following, based on an argument of Jensen.

7.5 *Theorem.* If a domain A satisfies (X) and (Y) and the quotient field Q of A is countably generated, then gl dim(A) = 2.

Proof: For $0 \neq a \in A$, tensor the presentation of I with A/aA to get

$$0 \longrightarrow L/aL \longrightarrow F/aF \longrightarrow I/aI \longrightarrow 0$$

Since L/aL is a flat A/aA-module, $\mathrm{pd}_{A/aA}(L/aL) < \infty$, and thus $\mathrm{pd}_{A/aA}(I/aI) \leq 1$. From the change of rings result of (5.6),

$$\mathrm{pd}_A(I/aI) \leq 2$$

Apply to the sequence

$$0 \longrightarrow I \xrightarrow{a} I \longrightarrow I/aI \longrightarrow 0$$

the functor $\mathrm{Hom}_A(-, M)$ to obtain the exact sequence

$$\mathrm{Ext}^2_A(I,M) \xrightarrow{a} \mathrm{Ext}^2_A(I,M) \longrightarrow \mathrm{Ext}^3_A(I/aI,M) = 0$$

It follows that $\mathrm{Ext}_A^2(I,M)$ is divisible by each nonzero element of A. To show that this module is zero, since Q is a countably generated module, it suffices to show that

$$\mathrm{Hom}_A(Q,\mathrm{Ext}_A^2(I,M) = 0$$

Consider the two spectral sequences with the same limit H^n:

$$E_2^{p,q} = \mathrm{Ext}_A^p(Q,\mathrm{Ext}_A^q(I,M)) \underset{p}{\Longrightarrow} H^n$$

$$\tilde{E}_2^{p,q} = \mathrm{Ext}_A^p(\mathrm{Tor}_q^A(Q,I),M) \underset{p}{\Longrightarrow} H^n$$

Since Q is A-flat, $\tilde{E}_2^{p,q} = 0$ for $q > 0$. Since $Q \otimes I = Q$, and Q is countably generated, $\mathrm{pd}(Q \otimes I) = 1$. Thus $\tilde{E}_2^{2,0} = 0$, and the two-dimensional cohomology group in the limit of the spectral sequence is zero. Considering the first spectral sequence, one has $E_2^{p,q} = 0$ for $p \geq 2$, since $\mathrm{pd}\ Q = 1$. This implies that $E_2^{0,2} = E_\infty^{0,2} = 0$ since $H^2 = 0$. But $E_2^{0,2} = \mathrm{Hom}_A(Q,\mathrm{Ext}_A^2(I,M))$ and $\mathrm{Ext}_A^2(I,M) = 0$ as desired.

2. *Symmetric algebras*

Symmetric algebras of flat modules provide interesting examples of rings of low global dimension. A number of difficulties arise when modules of rank two or higher are considered. Even estimates of its dimensions are hard to find and basic properties, *e.g.*, coherence, are missing in general. Here we concern ourselves mostly with rank one modules and Dedekind domains-precisely the case when rings of global dimension two appear.

7.6 *Lemma.* Let A be a GCD domain, and let M be a rank one flat A-module. Then M is the directed union of its cyclic submodules.

Proof: Assume that M is a submodule of K, the field of quotients of A. Let

$$\alpha = a'/b' \qquad \beta = c'/d'$$

be two elements of M, with $\mathrm{GCD}(a',b') = 1 = \mathrm{GCD}(c',d')$. Let $e = \mathrm{GCD}(a',c')$ $f = \mathrm{GCD}(b',d')$, and write

$$\alpha = ea/fb \qquad \beta = ec/fd$$

The relation

$$bc\alpha + (-ad)\beta = 0$$

yields, by the flatness of M,

$$\alpha = \Sigma b_n \alpha_n \qquad \beta = \Sigma c_n \alpha_n$$

$\alpha_n \in M$, $bcb_n = adc_n$. By our choice of a, we must have $b_n = ab'_n$, $c_n = cc'_n$. On the other hand, $bb'_n = dc'_n$ yields $b'_n = db''_n$, $c'_n = bc''_n$, $b''_n = c''_n$. Altogether we have $b_n = adb''_n$, $c_n = cbb''_n$. But then

$$\alpha = ad(\Sigma b''_n \alpha_n) \qquad \beta = cb(\Sigma b''_n \alpha_n)$$

7.7 *Proposition.* Let A be locally a GCD domain, and let M be a rank noe flat module. Then

$$\text{Tor-dim}(B) \leq 1 + \text{Tor-dim}(A)$$

where B is the symmetric algebra $S_A(M)$.

Proof: We may assume that A is a local ring. From (7.6), M is the directed union $\cup M_\alpha$ of cyclic submodules M_α. From the properties of symmetric algebras we get

$$S_A(M) = \varinjlim S_A(M_\alpha) = \varinjlim A[x_\alpha]$$

By the Syzygies' theorem the Tor-dimension of $A[x_\alpha]$ is $1 + \text{Tor-dim}(A)$, and the conclusion follows from Ref. 14.

Remark. Proposition 7.7 applies to the case when A is a coherent ring of finite Tor-dimension since they are locally GCD domains. It is likely that the Tor-dimension of $S_A(M)$ is at least Tor-dim(A).

Let A be a Prüfer domain and, let $0 \neq M$ be a submodule of K, the field of quotients of A.

7.8 *Proposition.* $B = S_A(M)$ satisfies:

(1) Tor-dim(B) = 1 if and only if $M = K$.

(2) If A is a Dedekind domain, B is a coherent ring.

Proof: B may be viewed as the following subring of K[t]

$$B = A + Mt + M^2t^2 + \cdots$$

wnere M^n denotes the submodule of K generated by products of n elements in M. If B has Tor-dimension one, for each $0 \neq a \in A$ the ideal

$$I = (a, Mt)$$

is flat and properly contains the prime ideal $P = (Mt)$.

7.9 *Lemma.* If a flat ideal I of a ring A properly contains a prime ideal P, then $P = PI$.

Proof: I/PI is a flat A/P-module and thus torsion-free. Tensor by A/P the sequence

$$0 \longrightarrow I \longrightarrow A \longrightarrow A/I \longrightarrow 0$$

to get

$$0 \longrightarrow \mathrm{Tor}_1^A(A/P,A/I) \longrightarrow I/PI \longrightarrow A/P \longrightarrow A/I \longrightarrow 0$$

But the submodule $\mathrm{Tor}_A^1(A/P,A/I)$ of I/PI is annihilated by I/P and is thus trivial, that is, $P \cap I = PI$, as desired.

Applying this lemma, we complete the proof of part (1) of (7.8) since it shows $M = aM$, and M is divisible.

When M = K, it is obvious that B is a Prüfer domain.

To show the coherence of B, let I be a finitely generated ideal. Since M is the directed union of its finitely generated submocules, B is the directed union of its subrings $B_\alpha = S_A(M_\alpha)$. Thus we may take $I = BI_0$, where I_0 is a finitely generated ideal in $B_0 = S_A(M_0)$. B_0 is however a Noetherian ring of global dimension two, and so we may write $I_0 = JL$, where J is a projective ideal and L is overdense ideal. It clearly suffices to show that LB has a finite presentation.

L being an ideal of grade 2 in B_0 (= locally at the primes of A a polynomial ring in one indeterminate over A), we must have $L \cap A = P \neq 0$. In the sequence

$$0 \longrightarrow PB \longrightarrow LB \longrightarrow LB/PB \longrightarrow 0$$

LB/PB is a finitely generated module over $C = B/PB = S_{A/P}(M/PM)$. Since A/P is an Artinian ring, M/PM is a finitely generated A/P-module. Thus LB/PB is a finitely presented module over C--in turn finitely presented over B. With both PB and LB/PB finitely presented, LB is finitely presented.

Remark. The coherence of the symmetric algebra of higher rank modules is not assured. Estimates of its dimensions are somewhat easier to give as in the following example. Let $A = \mathbb{Z}$ and $M = Q \oplus Q$, and write

$$B = S_{\mathbb{Z}}(Q^2)$$

B is the subring of Q[x,y] with integers for constant terms

$$B = \mathbb{Z} + Qx + Qy + \cdots$$

It is clear that $By{:}x = (Qy)$, which is a non-finitely generated ideal; thus B is not coherent.

The determination of the dimensions of B can be done in the following manner: Write $M = Q \oplus Q = Q \oplus \varinjlim G_\alpha$, G_α finitely generated. Then

$$S_{\mathbb{Z}}(M) = \varinjlim S_{\mathbb{Z}}(Q)[x_\alpha]$$

and $\text{Tor-dim}(S_{\mathbb{Z}}(Q^2)) \leq 2$. Since $S_{\mathbb{Z}}(M) \otimes Q = Q[x,y]$, we have that $\text{Tor-dim}(B) = 2$. Its global dimension is 3 since by Lemma 0.11, it is bounded by 3; and since it is a noncoherent domain, we have

$$\text{gl dim}(S_{\mathbb{Z}}(Q^2)) = 3.$$

We now embark on the determination of the global dimension of $B = S_A(M)$, where M is a rank one flat module over the Dedekind domain A. Note that although we know the Tor-dimension of B, lacking more information on the cardinality of M, it is not assured that B has finite global dimension. The procedure we follow was jointly worked out with J. Carrig.

Let X be the set of maximal ideals P of A with $M = PM$, and write Y for the primes where $M \neq PM$. Thus X consists of the primes P for which $M_P = K$ = field of quotients of A, whereas at a prime P of Y, $M_P \simeq A_P$. Let

$$A_X = \text{subring of } K \text{ generated by all } P^{-1},\ P \in X$$

$$A_Y = \text{subring of } K \text{ generated by all } P^{-1},\ P \in Y$$

A_X and A_Y are Dedekind domains, and the change of rings

$$A \longrightarrow A_X \times A_Y$$

is faithfully flat. By the descent result of Ref. 26, it suffices to prove that

$$S_{A_X}(M \otimes A_X) \quad \text{and} \quad S_{A_Y}(M \otimes A_Y)$$

both have global dimension at most two. Notice that $M \otimes A_Y = K$ and that at the primes of A_X, $M \otimes A_X$ is locally principal.

We first take care of B_Y. We note that B_Y fits in the diagram

$$\begin{array}{ccc} B_Y & \longrightarrow & K[t] \\ \downarrow & & \downarrow \\ A_Y & \longrightarrow & K \end{array}$$

This is a cartesian square close to the kind considered in Theorem 4.23; we conclude, as we did in (4.23) that $\text{gl dim}(B_Y) = 2$.

Assume now that M is a module locally free over A and let I be an ideal of B. If I is finitely generated, since Tor-dim(B) = 2, from (7.7) we conclude $\mathrm{pd}_B I \leq 1$. We also view B as embedded in K[t].

Let J be the ideal of B generated by the elements in I of degree r, where r = degree of polynomial generating IK in K[t]. We show that J is a flat ideal of projective dimension at most one and then study I/J. The statement about the flatness being clear (as at the localizations of A, B is locally A[t]), we proceed in a slightly broader context.

More generally, let A be a coherent domain of finite global dimension, and consider $B = S_A(M)$, for a flat A-module M. Assume that B has finite Tor-dimension and is coherent.

7.10 *Proposition.* Let I be a flat ideal of B. Then

$$\mathrm{pd}_B I \leq \mathrm{gl\ dim}(A)$$

The proof will consist of a number of observations on flat ideals of coherent rings.

Let I be a flat ideal of B. If $a_1, \ldots, a_n$ are elements of I, the ideal J they generate admits a decomposition

$$J = MN$$

with M projective and N overdense. We claim that $M \subset I$, and as a consequence I will be a directed union of projective ideals.

7.11 *Proposition.* Let I be a flat ideal of a commutative ring A, and a an element of I. If J is a finitely generated ideal containing a regular element with $J^{-1} = A$, then $Ja \subset I$ implies $a \in I$.

Proof: Take $j_1, \ldots, j_r$ generators of J, with j_1 a regular element. Then, by the flatness of I,

$$j_k(j_1 a) - j_1(j_k a) = 0$$

for $2 \leq k \leq r$ implies that there are finite sets of elements $c_{kn}, d_{kn} \in A$ and $\alpha_{kn} \in I$ with

$$j_1 a = \Sigma c_{kn\ kn} \quad \text{and} \quad j_k a = \Sigma d_{kn\ kn}$$

with $j_k c_{kn} = j_1 d_{kn}$ for all n. Thus $c_{kn} \in ((j_1):j_k$ and $j_1 a \in ((j_1):j_k)I$. Since this is true for all k, we have

$$j_1 a \in \bigcap_k [((j_1):j_k)I] = [\bigcap_k ((j_1):j_k)]I = ((j_1):J)I = j_1 I$$

since $J^{-1} = A$. Hence $a \in I$.

Proof of (7.10): Let I be a flat ideal of $B = S_A(M)$. IK is an ideal of BK generated by a single polynomial f that may be taken in I. View I as the directed union of all its (finitely generated) projective subideals containing f. For any such ideal P, fP^{-1} is an invertible ideal of B that we claim is generated by $fP^{-1} \cap A$. It is clear that $L_{(P)} = fP^{-1} \cap A$ is nonzero; that $L_{(P)}$ generates fP^{-1} follows since A is locally a GCD domain.

Write now

$$I = \varinjlim P = \varinjlim (Pf^{-1})f$$

Here $P^{-1}f = (L_{(P)})^{-1}B$, with $(L_{(P)})^{-1}$ a flat submodule of K. Thus we conclude

$$\mathrm{pd}_B I \leq \mathrm{gl\ dim}(A)$$

In fact we proved that the projective dimension of the field of quotients of A yields a bound for the projective dimension of flat ideals of B.

We now prove the main result of this section in the following theorem.

7.12 *Theorem*. Let M be a torsion-free, rank one module over the Dedekind domain A. Then $S_A(M)$ has global dimension two.

Proof: We may assume that M is locally finitely generated as an A-module. Let J be an ideal of $B = S_A(M)$, and denote by I the ideal generated by the elements of minimal "degree" of J. To show $\mathrm{pd}_B J \leq 1$, it suffices by (7.10) to show that $\mathrm{pd}_B(J/I) \leq 1$.

J/I as an A-module is a torsion module. It decomposes then into

$$J/I = \oplus\Sigma(J/I)^{(P)}$$

the P-primary components, P a prime of A. This decomposition is clearly a direct sum of B-modules and $(J/I)^{(P)} = (J/I)_P$, the localization of the module J/I at P. It suffices then to show that $\mathrm{pd}_B(J/I)^{(P)} \leq 1$.

Note that J_P is a finitely generated ideal of $B_P \simeq A_P[x]$, thus, as a B_P-module, $(J/I)_P$ has projective dimension at most one. Since $(J/I)_P$ is also a P-torsion B_P-module, let $0 \neq d$ be an element of P such that $d(J/I)_P = 0$. Let $P = P_1, \ldots, P_n$ be the associated primes of (d), and consider the change of rings

$$\phi: A \longrightarrow A_d \times A_{P_1} \times \cdots \times A_{P_n} = A'$$

Here A_d is the ring of fractions of A with respect to the multiplicative set of powers of d. ϕ is clearly faithfully flat. Let us examine

$$(J/I)_P \otimes_B B'$$

where

$$\phi_B: B \longrightarrow B_d \times B_{P_1} \times \cdots \times B_{P_n} = B'$$

But

$$(J/I)_P \otimes B' = (0, (J/I)_P, 0, \ldots, 0)$$

From Ref. 26 we conclude that $\mathrm{pd}_B(J/I)_P \leq 1$, and the proof of (7.12) is complete.

7.13 *Remarks*. It is clear that the converse of Theorem 7.12 is false. Symmetric algebras of modules over von Neumann rings provide examples where it is not difficult to estimate the dimensions. Thus, if A is a countable von Neumann ring and M is a countable A-module locally generated by 1 or 2 elements, then $S_A(M)$ has global dimension at most two. If one excludes these cases as well as those where A is a domain and M admits lots of divisibility, then (7.12) characterizes the symmetric algebras of global dimension two.

Another source of examples of rings of low dimension is provided by group algebras. However, when the coefficient ring is not a field, with the exception of some well-known algebras, they are hardly ever of global dimension two.

Let A be a ring, and let G be an abelian group. The group algebra A[G] may be viewed as $A[\varinjlim G_\alpha] = \varinjlim A[G_\alpha]$, G_α running over the finitely generated subgroups of G. This presentation yields:

(1) If A is a Noetherian ring, then A[G] is a coherent ring, since A[G] is a limit of Noetherian rings $A[G_\alpha]$ and $A[G_\beta]$ is $A[G_\alpha]$-flat when $G_\alpha \subset G_\beta$.

(2) If G is a torsion-free group, A[G] is the limit of rings isomorphic to $A[x_1, x_1^{-1}, \ldots, x_n, x_n^{-1}]$, n the rank of G. Thus the Tor-dimension of A[G] can be estimated as

$$\text{Tor-dim}(A[G]) \leq \text{rank}(G) + \text{Tor-dim}(A)$$

If A is a Noetherian ring, we can easily conclude that

$$\text{Tor-dim}(A[G]) = \text{rank}(G) + \text{gl dim}(A)$$

$$\text{gl dim}(A[G]) = \text{rank}(G) + \text{gl dim}(A) + \delta$$

where $\delta = 0, 1$ according to whether G is finitely generated.

The estimates are immediate consequences of the fact that $\mathbb{Z}^n$ embeds in G and thus the flatness of A[G] over $A[\mathbb{Z}^n]$ can be used to push across to A[G] a regular sequence of the appropriate size. Indeed, letting $a_1, \ldots, a_r$ be a regular sequence in A, $a_1, \ldots, a_r, x_1 - 1, \ldots, x_n - 1$ is a regular sequence in A[G]. From the change of rings theorem we conclude that

$$\text{Tor-dim}(A[G]) \geq r + n$$

and the first equality ensues.

As for the other equality, from Lemma 0.11 we already have

$$\text{gl dim}(A[G]) \leq \text{Tor-dim}(A[G]) + 1$$

That we have gl dim(A[G]) = Tor-dim(A[G]) only when G is finitely generated (G always of finite rank) follows from the coherence of A[G]. For if r above is gl dim(A), then FPD(B) = 0, where

$$B = A[G]/(a_1, \ldots, a_r, x_1 - 1, \ldots, x_n - 1)$$

Since B is coherent it must also be Artinian. But then G must be finitely generated.

3. Sheaf representation

Let A be a ring of global dimension two. Since A is locally a domain, there is a map

$$\phi\colon \text{Spec}(A) \longrightarrow \text{Min}(A)$$

that associates the unique minimal prime ideal $\phi(P)$ contained in P to the prime ideal P of A.

7.14 *Lemma.* If A is coherent, ϕ is continuous.

Proof: If $Q \in \text{Min}(A)$ and $f \in A \smallsetminus Q$, let $\overline{D}(f)$ be the open set of Min(A) defined as $\overline{D}(f) = \text{Min}(A) \cap D(f)$. Notice that by Proposition 6.1, $\overline{D}(f) = \overline{D}(e)$, e an idempotent of A. It is then clear that $\phi^{-1}(\overline{D}(e)) = D(e)$.

If $\mathcal{A}$ is the sheaf the ring A induces on Spec(A), denote by $\mathcal{B}$ the direct image of $\mathcal{A}$ under ϕ. It follows immediately that the fiber of $\mathcal{B}$ at $Q \in \text{Min}(A)$ is the domain A/Q.

Another sheaf carried by the minimal spectrum of a coherent ring of global dimension two is $\mathcal{C}$, obtained by restriction of $\mathcal{A}$ to Min(A). In this case $\mathcal{C}$ is just the structure sheaf associated to K, the total ring of quotients of A, which is a von Neumann ring (see Theorem 1.8).

Between themselves, $\mathcal{B}$ and $\mathcal{C}$ should provide various characterizations of A. It is not clear at the moment, however, how this kind of formulation will further elucidate the theory of the coherent rings of dimension two. As a problem, certainly the relations between the sheaf cohomology groups and group extensions over A should be studied [50].

Chapter 8

COHERENCE OF POLYNOMIAL RINGS

In this chapter we prove the stable coherence of the coherent rings of low homological dimensions promised in the Introduction. In too many ways the coherent rings differ from Noetherian rings, most strikingly in the failure of the "basis theorem": If a ring A is coherent, then A[x], the polynomial ring in one indeterminate over A, is not necessarily coherent [58]. Here we show, however, that if the deep structure of the coherent ring is "sufficiently" rich, then the coherence of the polynomial ring over A ensues. We discuss the subject of the λ-dimension of a ring in the last section of this chapter. Its scope is far wider than the rings of global dimension two, but the results are of a fragmentary nature so far. It appears to pose a frame where notions such as stable coherence can be more conveniently discussed.

1. A remark of Gruson's

Let A be a commutative ring, B an A-algebra of finite presentation, and let E be a finitely generated B-module. As in Ref. 26 we attempt to determine the nature of U, the set of points in Spec(B) where E is A-flat, and whether E admits a finite B-presentation over U. The following modification of Ref. 26 (3.4.6 of part I) was given by Gruson.

8.1 *Proposition.* Let A, B, and E be as above and let $h: A \to C$ be an injective ring homomorphism. Assume $E_C = E \otimes_A C$ is of finite presentation over $B_C = B \otimes_A C$. Then the set U of points of Spec(B), where E is A-flat, is open and E is of finite presentation over U.

Proof: Given a point P of U, let us find a neighborhood of P where the assertion is valid. By restricting Spec(B) as in Ref. 26 we can find a surjection of B-modules

$$f: F \to E$$

where F is of finite presentation and A-flat and f_P is an isomorphism.

Since F_C is of finite presentation we may assume, restricting Spec(B) still further if necessary, that $f \otimes_A C$ is an isomorphism. We deduce the assertion from the commutative diagram

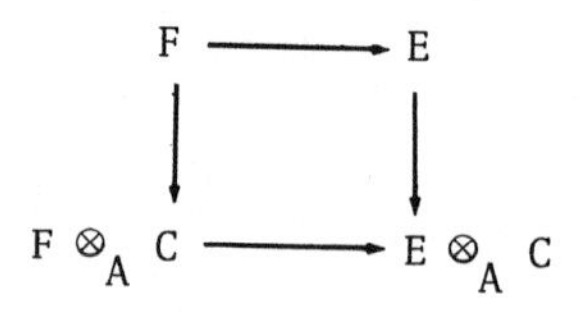

where the left vertical map is injective by the flatness of F.

We look at some consequences for stable (*i.e.*, polynomial) coherence. In all cases we consider, A is a ring whose total rings of quotients K is a von Neumann ring.

(a) Let A be a von Neumann ring. Let I be a finitely generated ideal of $B = A[x_1,\ldots,x_n]$. Then $E = B/I$ is a B-module of finite presentation, and E and B are A-flat. That I is finitely presented follows from the "principle of finite infinite presentation" (see 11.3.9.1 of Ref. 24, and also Ref. 3):

8.2 *Proposition*. Let B be an A-algebra of finite presentation, and let E be a B-module of finite presentation. Write

$$0 \longrightarrow L \longrightarrow B^n \longrightarrow E \longrightarrow 0$$

If B and E are A-flat, then L is a finitely presented B-module.

(b) A is semihereditary. If $B = A[x_1,\ldots,x_n]$ and I is a finitely generated ideal of B, I is A-flat. Takind $C = K$ in (8.1), we conclude that I is finitely presented.

Thus far we have shown that coherent rings of Tor-dimension less than or equal to one are stably coherent. The next order of complexity will be taken up in Section 2.

2. *Rings of global dimension two*

Let A be a ring of global dimension two having total ring of quotients K. If A is coherent, K is a von Neumann ring (Proposition 6.1). We will show that $B = A[x_1,\ldots,x_n]$ is coherent, by first reducing to the local case and then giving a proof in that situation using the structure theorem (2.2).

(1) *Reduction to the local case.* Let B and I be as above. If P is a

prime ideal of B where I is not A-flat, then $\underline{p} = P \cap A$ is such that $A_{\underline{p}}$ is not a valuation ring, and $\underline{p}$ is a maximal ideal of A and according to Theorem 6.6 finitely generated. Thus V, the set of prime ideals of B where I is not A-flat, is a closed set in Spec(B) by (8.1), and the primes in V are finitely generated. V is then a Noetherian space and thus the union of finitely many irreducible components,

$$V = V(P_1) \cup \cdots \cup V(P_n)$$

Let $\underline{p}_i = P_i \cap A$ and let $J = \underline{p}_1 \cdots \underline{p}_n \subset A$. Now J is finitely generated, say $J = (f_1,\ldots,f_m)$ (actually J can be generated by three elements). Consider the change of rings determined by the standard map

$$A \longrightarrow A_{\underline{p}_1} \times \cdots \times A_{\underline{p}_n} \times A_{f_1} \times \cdots \times A_{f_m}$$

This map is faithfully flat and in a component such as A_{f_i}, I_{f_i} is A-flat (and of finite presentation).

To complete the proof of the coherence of B, it suffices to see that $I \otimes_A A_{\underline{p}}$ is finitely presented over $A_{\underline{p}}[x_1,\ldots,x_n]$.

(2) *The local case.* Before focusing our attention on the coherence of the polynomial ring over a local ring of global dimension two, we will prove a proposition concerning coherence in cartesian squares. Then drawing upon the structure theory of local rings of global dimension two, we will apply this result to the question at hand.

Let

$$\begin{array}{ccc} R & \xrightarrow{i_1} & S \\ {\scriptstyle i_2}\downarrow & & \downarrow{\scriptstyle j_1} \\ T & \xrightarrow[j_2]{} & D \end{array}$$

be a cartesian square of commutative rings. Let j_1 and i_2 be surjective, let i_1 and j_2 be injective, and identify $H = \ker(i_2)$ with an ideal in S, that is, $H = HS$. Assume further that H is a flat ideal of R and S is flat as an R-module.

8.3 *Proposition.* Adopting the format above, if S is coherent and T is Noetherian, then R is coherent.

Proof: Consider the usual ring homomorphism $R \longrightarrow T \times S$ and note that the image of H is $(0,H)$: thus H is an ideal common to R and $T \times S$.

Let $L = (a_1,\ldots,a_n)$ be a finitely generated ideal in R, and consider an exact sequence

$$(*) \qquad 0 \longrightarrow K \longrightarrow R^n \longrightarrow L \longrightarrow 0$$

Tensoring (*) with S we see that $K \otimes S$ is finitely generated as an S-module since $L \otimes S$ is a finitely generated ideal of S and S is coherent and flat as an R-module.

Tensoring (*) with T, we have

$$0 \longrightarrow K \otimes T \longrightarrow T^n \longrightarrow L \otimes T \longrightarrow 0$$

which is also exact since $\mathrm{Tor}_1^R(R/H,L) = 0$, because H is a flat ideal and L is a submodule of a free R-module. Since T is Noetherian, $K \otimes T$ is a finitely generated T-module, and so $K \otimes (T \times S)$ is a finitely generated $(T \times S)$-module. However Theorem 3.1 provides the descent of finite generation applicable. Thus K is finitely generated, and so R is coherent.

As shown in Corollary 4.19, if A is a local ring of global dimension two, then A can be embedded in a cartesian square

$$\begin{array}{ccc} A & \xrightarrow{i_1} & A_P \\ {\scriptstyle i_2}\downarrow & & \downarrow \\ A/P & \longrightarrow & A_P/P \end{array}$$

where P is a flat prime ideal of A, A_P is a valuation domain, and A/P is Noetherian.

<u>8.4 *Proposition*</u>. If A is a local ring of global dimension two, then $A[x_1,\ldots,x_n]$ (=A[X]) is coherent.

<u>*Proof:*</u> If we tensor the cartesian square above with A[X], we get the cartesian square

$$\begin{array}{ccc} A[X] & \xrightarrow{i_1 \otimes 1} & A_P[X] \\ {\scriptstyle i_2 \otimes 1}\downarrow & & \downarrow \\ A/P[X] & \longrightarrow & A_P/P[X] \end{array}$$

where $\ker(i_2 \otimes 1) = P[X]$ and P[X] is a flat prime ideal in A[X].

According to Ref. 26, $A_P[X]$ is coherent and since A/P[X] is Noetherian, A[X] is coherent by the proposition above.

We sum up the results thus far in the following theorem.

<u>8.5 *Theorem*</u>. If A is a coherent ring of global dimension two, then $A[x_1,\ldots,x_n]$ is coherent.

8.6 *Remark.* Unpublished results of Gruson have considerably extended (8.5) to some coherent local rings of Tor-dimension two. An elementary but useful remark that he observed reduces the question of coherence in a polynomial ring over a local ring to the consideration of the special fiber.

Let A be a local ring and let M be its maximal ideal. Write $A(x) = A[x]_{M[x]} = A[x,S^{-1}]$ where $S = A[x] \setminus M[x]$. Then $A[x]$ is coherent if and only if $A(x)$ is coherent.

For a proof, let I be a set, and consider the cokernel C of

$$(A[x])^I \longrightarrow (A(x))^I$$

If A is coherent, C is A-flat because $A[x] \longrightarrow A(x)$ is universally injective. The conclusion follows from Theorem 0.14.

3. The λ-dimension of a ring

The discussion that follows is partly intended to provide a setting where the "Hilbert basis theorem" is "valid" more often.

Let A be a ring and let E be an A-module.

8.7 *Definition.* E is said to be of finite n-presentation if there is an exact sequence

$$F_n \longrightarrow F_{n-1} \longrightarrow \cdots \longrightarrow F_1 \longrightarrow F_0 \longrightarrow E \longrightarrow 0$$

with the F_i's free modules of finite rank.

We recall the familiar statement that follows.

8.8 *Proposition* (Schanuel's lemma). Let

$$0 \longrightarrow M \longrightarrow P_n \longrightarrow \cdots \longrightarrow P_0 \longrightarrow E \longrightarrow 0$$

and

$$0 \longrightarrow N \longrightarrow Q_n \longrightarrow \cdots \longrightarrow Q_0 \longrightarrow E \longrightarrow 0$$

be exact sequences, with the P_i's and Q_i's projective A-modules. Then

$$M \oplus Q_n \oplus P_{n-1} \oplus \cdots \simeq N \oplus P_n \oplus Q_{n-1} \oplus \cdots$$

8.9 *Definition.* Let E be an A-module. Write

$\lambda_A(E) = \sup\{n \mid \text{there is a finite n-presentation of E}\}$

Remark. With Ref. 11 (Chap. I) we also put $\lambda_A(E) = -1$, if E is not finitely generated. The Schanuel's lemma says here that given an A-module E, any finite n-presentation of E can be continued to a finite $\lambda_A(E)$-presentation.

We come now to our basic definition.

8.10 *Definition*. The λ-dimension of a ring A [λ-dim(A) for short] is the least integer r (or ∞ if none such exists) such that $\lambda_A(E) \geq r$ implies $\lambda_A(E) = \infty$.

Remark. B. Mitchell pointed out that this notion has also been referred to as r-coherence. A discussion of this λ-function is in Ref. 11, Chap I. (Exercise 8 there is in error; among other consequences it would make every domain a coherent ring.)

8.11 *Examples*. (1) λ-dim(A) = 0 iff A is Noetherian.

(2) λ-dim(A) = 1. For any module A/I, I a finitely generated ideal, $\lambda_A(A/I) = \infty$; in particular I is finitely generated. Thus A is coherent; the converse is well known.

(3) $A = \Pi C[x,y,z]$, a (infinite) countable product of copies of the polynomial ring C[x,y,z] (C = complex numbers). We show that λ-dim(A) = 2, using an argument of Quentel. According to the Macaulay's examples [1], in the ring C[x,y,z] there is no bound for the number of elements needed to generate height two prime ideals. Pick then a sequence of prime ideals P_1, P_2, ... with a strictly increasing number of elements for minimum generating sets. Each P_i is minimal over $(f_{1,i},f_{2,i})$ and can be written

$$P_i = (f_{1,i},f_{2,i}):g_i$$

Consider the elements of A:

$$F_1 = (f_{1,i}) \qquad F_2 = (f_{2,i}) \qquad G = (g_i)$$

Since $(F_1,F_2):G$ is not finitely generated, A is not coherent.

Now for the proof that λ-dim(A) $\leq$ 2: Let

$$0 \longrightarrow M \longrightarrow A^{r_2} \longrightarrow A^{r_1} \longrightarrow A^{r_0} \longrightarrow E \longrightarrow 0$$

be an exact sequence with r_0, r_1, and r_2 finite. We must show that M is finitely generated. If this sequence is read in the jth component of A, we see that M_j is C[x,y,z]-projective, of rank at most $r_2 - r_1 + r_0$. Therefore M is finitely generated.

(4) Let $B = \Pi Q[x,y]$ (infinite) countable product of copies of Q[x,y], the polynomial ring in two indeterminates over the rationals. B is easily seen (using the argument above) to be coherent. Soublin [58] has shown however that A = B[T], T an indeterminate, is not coherent. We shall soon see that λ-dim(A) = 2.

The following result is a crude but useful vehicle for finding first estimates of the λ-dimension of a ring A. We shall, in the sequel, disregard the case of a finite ring.

Let $P(A) = \prod_{\alpha\varepsilon I} A_\alpha$, $\text{card}(I) = \text{card}(A)$, $A_\alpha \simeq A$.

8.12 *Proposition.* Let E be a finitely presented A-module. Then $\lambda_A(E) \geq r > 1$ iff $\text{Tor}_i^A(P,E) = 0$ for $1 \leq i < r$.

Proof: There will be no harm in working with $r = 2$. Let

$$0 \longrightarrow L \longrightarrow F \longrightarrow E \longrightarrow 0$$

be a finite presentation of E. Suppose $\text{Tor}_1^A(P,E) = 0$, and let us show L to be finitely generated. See Ref. 40.

8.13 *Lemma.* Let L be a module of cardinality $\leq \text{card}(I)$. Then L is finitely presented if and only if the canonical mapping

$$\phi_L: P \otimes_A L = (\Pi A_\alpha) \otimes_A L \longrightarrow \Pi(A_\alpha \otimes_A L) = \Pi L_\alpha$$

given by $\phi_L((a_\alpha) \otimes m) = (a_\alpha \otimes m)$, is an isomorphism.

Proof: Suppose first that ϕ_L is surjective. Let $e = (e_\alpha) \varepsilon \Pi L_\alpha$, where we assume $\{e_\alpha\}$ includes all the elements of L at least once. If

$$e = \phi_L\left(\sum_{i=1}^{n} (a_\alpha)_i \otimes e_i\right)$$

then each e_α is a linear combination of the e_i's.

Write

$$0 \longrightarrow M \longrightarrow F \longrightarrow L \longrightarrow 0$$

with F free of finite rank. We have the commutative diagram

$$\begin{array}{ccccccccc} & & (\Pi A_\alpha) \otimes M & \longrightarrow & (\Pi A_\alpha) \otimes F & \longrightarrow & (\Pi A_\alpha) \otimes L & \longrightarrow & 0 \\ & & \downarrow{\scriptstyle \phi_M} & & \downarrow{\scriptstyle \phi_F} & & \downarrow{\scriptstyle \phi_L} & & \\ 0 & \longrightarrow & \Pi M_\alpha & \longrightarrow & \Pi F_\alpha & \longrightarrow & \Pi L_\alpha & \longrightarrow & 0 \end{array}$$

Thus ϕ_M is surjective if ϕ_L is bijective. According to the preceding, M is finitely generated.

The converse is just as easy.

8.14 *Corollary.* $\lambda\text{-dim}(A) \leq \mu + 1$, where $\mu = \text{flat dim}_A P$.

The preceding remarks can be used to derive estimates of the λ-dimensions of rings obtained by polynomial and power series extensions.

8.15 *Corollary*. Let A be a coherent ring, and let $B = A[T_1, \ldots, T_n]$. Then λ-dim(B) $\leq n + 1$.

Proof: Consider $P = P(B) = \Pi B_\alpha$ as above, noting that P is a flat A-module. By repeated use of the Syzygies' theorem, we conclude that flat $\dim_B P \leq n$, and so apply (8.14).

In general the estimate above is far too high. It is also not known whether λ-dim(B) $\leq n + \lambda$-dim(A), the form of the Hilbert's basis theorem desired.

8.16 *Example*. Let A be a noncoherent ring of Tor-dimension one (see Example 1.3b). λ-dim(A) = 2. In this case λ-dim(A[T]) = 2 also. Indeed, let

$$0 \longrightarrow L \longrightarrow F_1 \xrightarrow{\phi} F_0 \longrightarrow E \longrightarrow 0$$

be a finite presentation of E, $\lambda_B(E) \geq 2$ (B = A[T]). Since Tor-dim(A) = 1, Tor-dim(B) = 2; thus L is a finitely generated flat A[T]-module. We show that L is finitely presented, that is, projective, by utilizing a technical device suggested by S. Cox.

8.17 *Lemma*. Let

$$0 \longrightarrow L \xrightarrow{v} F \xrightarrow{u} G$$

be an exact sequence of finitely generated free modules over a commutative ring A. Then rank(F) = rank(L) + rank(u). (Here rank(u) means the last integer n for which the nth exterior power of u is nonvanishing.)

Proof: Tensor the exact sequence by A[T], without changing any of the ranks above. By McCoy's theorem [36] the ideal I(v) generated by the minors of order rank(L) of v is faithful. Since finitely generated faithful ideals in A[T] contain regular elements, choose f regular in I(v). Without affecting the ranks we may localize at the powers of f. The sequence splits then from the left, yielding the desired equality.

To continue with the proof of (8.16), since A is locally a domain, L is locally a free module. The rank of L is by the above locally constant in the Zariski topology [rank(ϕ) being so]. Thus by Ref. 11 L is projective.

The argument also shows that if A is a ring of Tor-dimension > 1, then λ-dim(A) $\leq$ Tor-dim(A).

For power series rings the comparison of the dimensions of A and B = A[[T]] is simpler:

8.18 *Proposition.* $\lambda\text{-dim}(B) \leq 1 + \lambda\text{-dim}(A)$.

Proof: Suppose $\lambda\text{-dim}(A) = r$ and, let E be a B-module, $\lambda_B(E) \geq r + 1$. Let

$$0 \longrightarrow L \longrightarrow F_{r+1} \longrightarrow F_r \longrightarrow \cdots \longrightarrow F_1 \xrightarrow{\phi} F_0 \longrightarrow E \longrightarrow 0$$

be a finite $(r + 1)$-presentation of E; we must show that L is finitely generated. Tensor the sequence by $B/(T)$ to get a finite r-presentation for the module $\text{im}(\phi)/T\text{im}(\phi)$. Since $\lambda\text{-dim}(A) = r$, L/TL is a finitely generated A-module. Since B is T-complete and L is a submodule of the free module F_{r+1}, L itself is finitely generated.

Let V be a non-Noetherian valuation domain. From (8.18), since V is coherent, $\lambda\text{-dim}(V[[T]]) \leq 2$. In fact, from Ref. 33 we have the following proposition.

8.19 *Proposition.* If $\text{rank}(V) > 1$, $\lambda\text{-dim}(V[[T]]) = 2$.

Proof: If M denotes the maximal ideal of V, we have

$$\text{flat dim}_{V[[T]]}(V[[T]]/(M,T)) = 2$$

If V[[T]] is coherent, its Tor-dimension will be two. By Theorem 5.21, V[[T]] is then integrally closed, which only happens with $\text{rank}(V) = 1$ according to Ref. 11.

The case $\text{rank}(V) = 1$ is a bit more delicate. Gruson has proved that in this case V[[T]] is coherent. More interesting is his claim that if the value group of V is the real numbers, then V[[T]][X] is not coherent.

Problem. Exhibit all positive integers as λ-dimensions of commutative rings.

4. Stability

In this section we use some of the preceding to obtain estimates for the λ-dimension of $B = A[T_1,\ldots,T_n]$, with A = regular. "Regular" here means that $r = \text{Tor-dim}(A)$ is finite. A more natural definition would require that finitely generated ideals have finite flat dimension; without putting a bound on these dimensions it is not clear that A[T] will inherit the property.

8.20 *Proposition.* (1) $\lambda\text{-dim}(B) \leq 1 + r$.

(2) If A is a domain, $\lambda\text{-dim}(B) \leq r$.

(3) If A is coherent, $\lambda\text{-dim}(B) \leq \inf\{r, n + 1\}$.

Proof: Let E be a B-module with $\lambda_B(E) \geq r + 1$. Consider a finite $(r + 1)$-resolution

$$0 \longrightarrow M \longrightarrow F_r \xrightarrow{\phi_r} F_{r-1} \longrightarrow \cdots \longrightarrow F_0 \longrightarrow E \longrightarrow 0$$

We must show that M, which is finitely generated, is finitely presented. Since $\operatorname{im}(\phi_r)$ is A-flat and finitely generated over B, (8.2) implies (1).

(2) and (3) are consequences of (8.1). Let $\lambda_B(E) \geq r$; if K is the field of quotients of A in case (2), or the von Neumann ring that is the total ring of quotients of A in case (3) the change of rings $A \to K$ will work. Using (8.15), the proof of (3) is complete.

We introduces now a condition that leads, at times, to considerable improvement in the estimates above.

Condition (L_i). For every $\underline{p} \in \operatorname{Spec}(A)$, $\lambda\text{-dim}(B_{\underline{p}}) \leq i$.

8.21 *Theorem.* Let A be coherent and $1 < s = \operatorname{gl}\dim(A) < \infty$; then $\lambda\text{-dim}(B) \leq s - 1$, if B satisfies condition (L_{s-1}).

Proof: We use the notation of (8.20). Suppose that $\lambda_B(E) \geq s - 1$:

$$F_{s-1} \longrightarrow \operatorname{im}(\phi_{s-1}) = N \hookrightarrow F_{s-2} \cdots$$

Let V be the set of primes of B where N is not A-flat. For $P \in V$ let $\underline{p} = P \cap A$; if $N_{\underline{p}}$ is not A-flat, since N is an $(s - 1)$-syzygy module, $\text{Tor-dim}(A_{\underline{p}}) > s - 1$. Thus

$$\text{Tor-dim}(A_{\underline{p}}) = \operatorname{gl}\dim(A_{\underline{p}}) = s$$

A is then super-regular at $\underline{p}$ in the sense of Theorem 5.29. The argument there easily adapts to the nonlocal case and shows that $\underline{p}$ is a finitely generated maximal ideal. Therefore, as in (8.5), V is a Noetherian space, and

$$V = V(P_1) \cup \cdots \cup V(P_m)$$

If $\underline{p}_i = P_i \cap A$, a change of rings as in (8.5) reduces the problem to the consideration of $B_{\underline{p}_i}$. The proof will be complete once B satisfies condition (L_{s-1}).

Note that (8.4) is precisely the statement that if A is a local ring of global dimension two, then B satisfies (L_1).

A point that emerges from this discussion of polynomial coherence is that the indeterminates must be thrown in simultaneously. This fact has also been present in recent work of Dobbs and Papick and especially in Gruson's extension of stable coherence to some rings of Tor-dimension two.

They lend credence to the following.

Conjecture. If A[T] is coherent, then $A[T_1,\dots,T_n]$ is coherent.

Remark. A more general result on the "passage from the punctual to the global" has been sketched by Gruson and will appear in a thesis by B. Alfonsi. Specifically,

8.22 *Theorem.* Let A be a commutative ring satisfying that for every complex L,

$$0 \longrightarrow L_n \longrightarrow L_{n-1} \longrightarrow \cdots \longrightarrow L_0$$

of free A-modules of finite type, the set of points $\underline{p}$ of Spec(A) such that L is acyclic at $\underline{p}$ is open. Then for every A-algebra of finite presentation B and every complex P,

$$0 \longrightarrow P_n \longrightarrow P_{n-1} \longrightarrow \cdots \longrightarrow P_0$$

of B-modules of finite presentation and A-flat, the set of points P of Spec(B) such that P is acyclic at P is open.

REFERENCES

1. S. Abhyankar, On Macaulay's examples, pp. 1-16, Lecture Notes in Mathematics 311, Springer-Verlag, Berlin, 1972.
2. M. Auslander, On the dimension of modules and rings, II, Nagoya Math. J. 9 (1955), 67-77.
3. M. Auslander and D. Buchsbaum, Homological dimension in local rings, Trans. Amer. Math. Soc. 85 (1957), 390-405.
4. M. Auslander and D. Buchsbaum, Unique factorization in regular local rings, Proc. Nat. Acad. Sci. USA 45 (1959), 733-734.
5. M. Auslander and D. Buchsbaum, Homological dimension in Noetherian rings, II, Trans. Amer. Math. Soc. 88 (1958), 194-206.
6. H. Bass, Finitistic dimension and a homological generalization of semi-primary rings, Trans. Amer. Math. Soc. 95 (1960), 466-488.
7. H. Bass, K-Theory and Stable Algebra, Publ. Math. IHES 22, Paris, 1964.
8. G. Bergman, Hereditary commutative rings, and centers of hereditary rings, Proc. London Math. Soc. 23 (1971), 214-236.
9. J. Bertin, Anneaux cohérents réguliers, C. R. Acad. Sc. Paris 273 (1971), 1-2.
10. P. Le Bihan, Sur la cohérence des anneaux de dimension homologique deux, C. R. Acad. Sc. Paris 273 (1971), 342-345.
11. N. Bourbaki, Algèbre Commutative, Hermann, Paris, 1961-1965.
12. L. Burch, On ideals of finite homological dimension in local rings, Proc. Camb. Phil. Soc. 64 (1968), 941-948.
13. L. Burch, A note on the homology of ideals generated by three elements in local rings, Proc. Camb. Phil. Soc. 64 (1968), 949-952.
14. H. Cartan and S. Eilenberg, Homological Algebra, Princeton University Press, Princeton, N.J., 1956.
15. S. U. Chase, Direct product of modules, Trans. Amer. Math. Soc. 97 (1960), 457-473.
16. S. Cox and R. Pendleton, Rings for which certain modules are projective, Trans. Amer. Math. Soc. 150 (1970), 139-156.
17. P. M. Eakin, The converse to a well known theorem on Noetherian rings, Math. Annalen 177 (1968), 278-282.
18. S. Endo, On semihereditary rings, J. Math. Soc. Japan 13 (1961) 109-119.
19. D. Ferrand, Descent de la platitude par un homomorphisme fini, C. R. Acad. Sc. Paris 269 (1969), 946-949.
20. H. Flanders, Tensor and exterior powers, J. Algebra 7 (1967), 1-24.

21. O. Forster, Uber the Anzahl der Erzeugenden eines Ideals in einem Noetherschen Ring, Math. Zeit. 84 (1964), 80-89.

22. R. Gilmer and W. Heinzer, On the number of generators of an invertible ideal, J. Algebra 14 (1970), 139-159.

23. B. Greenberg, Global dimension of cartesian squares, J. Algebra 32 (1974), 31-43.

24. A. Grothendieck, Eléments de Géométrie Algébrique, Publ. Math. IHES 32, Paris, 1965.

25. L. Gruson, Dimension homologique des modules plats sur un anneau commutatif noethérien, in Symposia Mathematica XI, 243-254, Academic Press, London, 1973.

26. L. Gruson and M. Raynaud, Critères de platitude et de projectivité, Inventiones Math. 13 (1971), 1-89.

27. T. Gulliksen, Tout idéal premier d'un anneau noethérien est associé à un idéal engendré par trois éléments, C. R. Acad. Sc. Paris 271 (1970), 1206-1207.

28. R. Heitmann, Generating ideals in Prüfer domains, Pacific J. Math. (to appear).

29. C. U. Jensen, Homological dimensions of rings with countably generated ideals, Math. Scand. 18 (1966), 97-105.

30. C. U. Jensen, Homological dimensions of $\aleph_0$-coherent rings, Math. Scand. 20 (1967), 55-60.

31. C. U. Jensen, Remarks on a change of rings theorem, Math. Zeit. 106 (1968), 395-401.

32. C. U. Jensen, On the vanishing of $\varprojlim^{(i)}$, J. Algebra 15 (1970), 151-166.

33. S. Jøndrup and L. W. Small, Power series over coherent rings, Math. Scand. 35 (1974), 21-24.

34. I. Kaplansky, Projective modules, Annals of Math. 68 (1958), 372-377.

35. I. Kaplansky, Fields and Rings, University of Chicago Press, Chicago, 1969.

36. I. Kaplansky, Commutative Rings, Allyn and Bacon, Boston, 1971.

37. I. Kaplansky, The homological dimension of a quotient field, Nagoya Math. J. 27 (1966), 139-142.

38. P. Kohn, Ph.D. Thesis, University of Chicago, 1972.

39. D. Lazard, Autour de la platitude, Bull. Soc. Math. France 97 (1969), 81-128.

40. H. Lenzing, Endlich präsentierbare Moduln, Arch. Math. 20 (1969), 262-266.

41. R. MacRae, On the homological dimension of certain ideals, Proc. Amer. Math. Soc. 14 (1963), 746-750.

42. R. MacRae, On an application of the Fitting invariants, J. Algebra 2 (1965), 153-169.

43. J. Marot, Anneaux héréditaires commutatifs, C. R. Acad. Sc. Paris 269 (1969), 58-61.

44. J. Milnor, Introduction to Algebraic K-Theory, Princeton University Press, N.J., 1971.

45. K. Nagarajan, Groups acting on Noetherian rings, Nieuw Arch. v. Wiskunde 16 (1968), 25-29.

46. B. L. Osofsky, Global dimension of valuation rings, Trans. Amer. Math. Soc. 127 (1967), 136-149.

47. B. L. Osofsky, Upper bounds of homological dimensions, Nagoya Math. J. 32 (1968), 315-322.

48. B. L. Osofsky, Homological dimensions and the continuum hypothesis, Trans. Amer. Math. Soc. 132 (1968), 217-230.

49. B. L. Osofsky, A commutative local ring with finite global dimension and zero divisors, Trans. Amer. Math. Soc. 141 (1969), 377-385.

50. R. S. Pierce, Modules over commutative regular rings, Mem. Amer. Math. Soc. 70 (1967).

51. Y. Quentel, Sur la compacité du spectre minimal d'un anneau, Bull. Soc. Math. France 99 (1971), 265-272.

52. Y. Quentel, Sur le théoreme d'Auslander-Buchsbaum, Colloque d'Algèbre Commutative, Rennes, 1972.

53. G. Sabbagh, Coherence of polynomial rings and bounds in polynomial ideals, J. Algebra 31 (1974), 499-507.

54. J. D. Sally and W. V. Vasconcelos, Flat ideals, I, Communications in Algebra 3 (1975), 531-543.

55. P. Samuel, Lectures on Unique Factorization Domains, Tata Inst. of Fund. Res., Bombay, 1964.

56. J. P. Serre, Sur la dimension homologique des anneaux et des modules noethériens, Proc. Int. Symp. Tokyo-Nippo (1955), 175-189.

57. H. Sheng, Finiteness of prime ideals in rings of global dimension two, Proc. Amer. Math. Soc. 35 (1973), 381-386.

58. J. P. Soublin, Anneaux et modules cohérents, J. Algebra 15 (1970), 455-472.

59. R. G. Swan, The number of generators of a module, Math. Zeit. 102 (1967), 318-322.

60. W. V. Vasconcelos, On finitely generated flat modules, Trans. Amer. Math. Soc. 138 (1969), 505-512.

61. W. V. Vasconcelos, On projective modules of finite rank, Proc. Amer. Math. Soc. 22 (1969), 430-433.

62. W. V. Vasconcelos, Annihilators of modules with a finite free resolution, Proc. Amer. Math. Soc. 29 (1971), 440-442.

63. W. V. Vasconcelos, The local rings of global dimension two, Proc. Amer. Math. Soc. 35 (1972), 381-386.

64. W. V. Vasconcelos, Rings of global dimension two, pp. 243-251, Lecture Notes in Mathematics 311, Springer-Verlag, Berlin, 1972.

65. H. Weyl, David Hilbert and his mathematical work, Bull. Amer. Math. Soc. 50 (1944), 612-654.

INDEX